This book is a further contribution to the series *Cambridge Studies in Philosophy and Biology*. It is an ambitious attempt to explain the relationship between intelligence and environmental complexity, and in so doing to link philosophy of mind to more general issues about the relations between organisms and environments, and to the general pattern of "externalist" explanations.

Two sets of questions drive the argument. First, is it possible to develop an informative philosophical theory about the mind by linking it to properties of environmental complexity? Second, what is the nature of externalist patterns of explanation? What is at stake in attempting to understand the internal in terms of the external?

The author provides a biological approach to the investigation of mind and cognition in nature. In particular he explores the idea that the function of cognition is to enable agents to deal with environmental complexity. The history of the idea in the work of Dewey and Spencer is considered, as is the impact of recent evolutionary theory on our understanding of the place of mind in nature.

Complexity and the function of mind in nature

CAMBRIDGE STUDIES IN PHILOSOPHY AND BIOLOGY

General Editor
Michael Ruse University of Guelph

Advisory Board
Michael Donoghue *Harvard University*
Jonathan Hodge *University of Leeds*
Jane Maienschein *Arizona State University*
Jesus Mosterin *University of Barcelona*
Elliott Sober *University of Wisconsin*

This major series publishes the very best work in the philosophy of biology. Nonsectarian in character, the series extends across the broadest range of topics: evolutionary theory, population genetics, molecular biology, ecology, human biology, systematics, and more. A special welcome is given to contributions treating significant advances in biological theory and practice, such as those emerging from the Human Genome Project. At the same time, due emphasis is given to the historical context of the subject, and there is an important place for projects that support philosophical claims with historical case studies.

Books in the series are genuinely interdisciplinary, aimed at a broad cross-section of philosophers and biologists, as well as interested scholars in related disciplines. They include specialist monographs, collaborative volumes, and – in a few instances – selected papers by a single author.

Alfred I. Tauber *The Immune Self: Theory or Metaphor?*
Elliott Sober *From a Biological Point of View*
Robert N. Brandon *Concepts and Methods in Evolutionary Biology*

Complexity and the function of mind in nature

PETER GODFREY-SMITH

Stanford University

CAMBRIDGE
UNIVERSITY PRESS

PUBLISHED BY THE PRESS SYNDICATE OF THE UNIVERSITY OF CAMBRIDGE
The Pitt Building, Trumpington Street, Cambridge CB2 1RP

CAMBRIDGE UNIVERSITY PRESS
The Edinburgh Building, Cambridge CB2 2RU, United Kingdom
40 West 20th Street, New York, NY 10011-4211, USA
10 Stamford Road, Oakleigh, Melbourne 3166, Australia

First published 1996
First paperback edition 1998

Library of Congress Cataloging-in-Publication Data is available.

A catalog record for this book is available from the British Library.

ISBN   0-521-45166-3 hardback
ISBN   0-521-64624-3 paperback

Transferred to digital printing 2004

For Rory Donnellan and George Madarasz

Contents

Contents

Contents

Preface

Although this book was brewing for a number of years, it owes many aspects of its present form to a series of discussions with Richard Francis during 1992 and 1993. Through these discussions a vague plan to investigate the "function of mind in nature," and the relations between intelligence and environmental complexity, took on a more definite shape. It was also in these discussions that this set of questions in the philosophy of mind became linked to more general issues about the relations between organisms and environments, and the general pattern of "externalist" explanations – explanations of internal properties of organic systems in terms of external properties.

So the book is intended to address two types of questions at once. First, is it possible to develop an informative philosophical theory about the mind, by linking mind to properties of environmental complexity? The second set of questions concerns externalist patterns of explanation in general. What are these explanations like? What are the characteristic debates and issues that surround the attempt to understand the internal in terms of the external?

I see these questions as closely linked, and in this book they are often examined simultaneously. However, for practical reasons the book is divided into two parts which are largely self-contained. This division is partly methodological and partly a matter of content. Part I is intended as a self-contained essay in the philosophy of mind, and it is completely nontechnical. Part II is focused on some specific biological models, which treat properties such as adaptive plasticity and their relation to variability in the environment. Part II has some technical discussions, although it is intended to be comprehensible to people with just a very basic knowledge of evolutionary biology and mathematics. Part II could in principle be read by itself, but those who are interested in philosophy of biology, or philosophy of science more generally, may do better to read the first few sections of Chapter 1, then Chapters 2 and 5, and then Part II. Of course, my hope is that all the philosophers of mind who get to the end of Part I will surge on into Part II, and that the philosophers of science will find themselves reading the rest of Part I as well.

Although much of the book is the product of a particular set of discussions, this acknowledgement should not be taken to suggest that Richard Francis agrees with the positions advanced in this book. One of the useful things about these discussions was the fact that they took place between one person who has long felt the pull towards various externalist views, especially externalist doctrines about the mind, and another person (Francis) who is more constitutionally suspicious of this pattern of explanation. I am also grateful to Richard Francis for his incisive comments on drafts and many solutions to specific problems that arose along the way. In keeping with the focus on environments I should also acknowledge the favorable physical environment in which the original discussions took place – the old Encina Gym at Stanford.

This project has also been greatly influenced by the writings of Richard Lewontin, and by discussions and correspondence with him. That influence will be evident in nearly every chapter. In addition, this book is the product of ten years of thinking about Fred Dretske's work in epistemology and the philosophy of mind. In some respects this book is an attempt to knit together a set of themes from three very different writers: Dewey, Dretske and Lewontin.

Important parts of the treatment of externalist and internalist explanations in biology were worked out in discussions with Rasmus Winther, and these discussions have benefited at many points from his ideas and his scholarship. Other themes in the book derive from my PhD thesis, written at UCSD under the superlative supervision of Philip Kitcher. Key improvements were also made as a consequence of the constructive antipodean skepticism of Kim Sterelny and others in the philosophy department at Victoria University, New Zealand, when I visited there in 1994.

In my attempts to come to grips with the models discussed in Part II of the book, the expertise, advice and patience of Sally Otto were invaluable. The software package *Mathematica* was constantly useful in this part of the project as well. For help on biological matters I am also indebted to Aviv Bergman, Carl Bergstrom, Ric Charnov, Marc Feldman, Deborah Gordon, David Pollock, Jon Seger and Sonia Sultan. Remaining errors in these discussions are my own efforts.

Sally Otto, Tim Schroeder, Elliot Sober, Kim Sterelny, Stephen Stich and Rasmus Winther all wrote detailed comments on earlier drafts. Devin Muldoon assisted with the graphics. For other comments, correspondence and discussions, encouragement of various kinds, or for making exactly the right suggestion at the right time, I would like to thank Ron Amundson, Tom Burke, Fiona Cowie, Daniel Dennett, Michael Devitt, Fred Dretske, Stephen Downes, Michael Ghiselin, Stephen Gould, Lori Gruen, P. J. Ivanhoe, Philip Kitcher, Richard Levins, Elisabeth Lloyd, John McDermott, Ruth Millikan, Andrew Milne, Greg O'Hair, Robert Richards, Michael Ruse, Peter Todd, and Tom Wasow. Parts of this work have been supported by research grants from the Center for the Study of Language and Information, and the Office of Technology Licensing, Stanford University.

San Francisco, California
December 1994

PART I

Foundations

1

Naturalism and Teleology

1.1 Basics

This book is largely about a single idea concerning the place of mind within nature. The idea is this:

Environmental Complexity Thesis:
The function of cognition is to enable the agent to deal with environmental complexity.

Naturalistic philosophy has already developed part of a theory of the place of mind within nature, a physicalist theory of what minds are made up of, and of how some of the strange properties of mentality can exist in the natural world. It may be possible to also develop another kind of theory of the place of mind in nature, a theory of what mind is *doing* here, perhaps a theory of what it is *for*. The environmental complexity thesis expresses one possible way to develop such a theory.

The topic of this book lies at the intersection of philosophy of mind, philosophy of biology, and epistemology. The aim is an account of the place of mind in nature, but many of the concepts used will be biological. And although I will not give a "theory of knowledge," this book is intended as a contribution to epistemology. In 1976 Alvin Goldman motivated an approach to epistemology, known as "reliabilism," by appealing to a sense of the word "know" illustrated by a phrase from Shakespeare's *Hamlet*: "I know a hawk from a handsaw." The present work is intended to shed light upon another sense of "know." This sense can also be illustrated with some well-known lines of verse:

> You've got to know when to hold 'em; know when to fold 'em.
> Know when to walk away; know when to run.
> "The Gambler," recorded by Kenny Rogers

3

A philosophical term standardly associated with epistemology which addresses this relation between thought and action is "pragmatism." This book aims to develop a theory which is pragmatist in some senses, but which is also closely allied to contemporary naturalistic views, including reliabilism.

Some parts of this book will investigate the environmental complexity thesis directly. Other parts will try to shed light on it from a distance, by examining a more general family of ideas which the thesis belongs to.

The environmental complexity thesis asserts a link between certain capacities of organic systems and a specific property of these systems' environments. It tries to understand the internal in terms of relations to the external. In this book all explanations of internal properties of organic systems in terms of external properties will be called *externalist*. Externalism is one very basic approach that can be taken to understanding the organic world, and one way the environmental complexity thesis will be investigated in this book is by means of a general investigation of the logic and pattern of externalist explanation.

There is also a subcategory of externalist explanation, exemplified by the environmental complexity thesis, which we will pay particular attention to. These are explanations in which properties of complexity are the focus: the internal complexity of some organic system is explained in terms of the complexity of its environment. These will be called "c-externalist" explanations.

The aim is not to discuss the categories of externalist and c-externalist explanation in order to promote them. They need no promotion. Externalism is one of a small number of basic explanatory schemas, encountered constantly within different parts of science and also in philosophy. The logic of externalist explanation is the logic of adaptationist evolutionary thought, associationist psychology such as behaviorist learning theory, and many brands of empiricist epistemology. It is an explanatory project which also prompts a distinctive pattern of revolt, advocating the explanatory importance of the internal structure of organic systems. This internalist response is exemplified in biology by developmentally oriented views of evolution, Chomsky's "mentalism" in linguistics and psycholinguistics, and various types of philosophical rationalism. One aim of this book is to locate the place occupied by the environmental complexity thesis within a very large conceptual landscape.

In the case of the general categories of externalist and c-externalist explanation, my aim is understanding rather than promotion. The more specific view, the environmental complexity thesis about cognition, *will* be promoted though. Once it has been clarified and refined in the right way, this idea has promise as a central component of a theory of mind's place in nature.

4

1.2 Spencer and Dewey

One way the environmental complexity thesis will be investigated is through an investigation of externalism in general. Another way the thesis will be investigated is historical. We will look in detail at two very different versions of the thesis, found in the work of Herbert Spencer and John Dewey.

Herbert Spencer is not often remembered now, and when he is rescued from oblivion it is usually to briefly damn him further. But he was an important intellectual figure during the Victorian period. He is interesting as an early and systematic exponent of a theory of mind based upon a naturalistic outlook and a set of evolutionary concepts involving adaptive adjustment to complex environments. Spencer published the first edition of his evolutionary *Principles of Psychology* in 1855, four years before Darwin's *Origin of Species*. In this work Spencer stressed the continuity between life and mind, which were understood as the "adjustment of internal relations to external relations." Spencer viewed mind as an advanced form of living organization developed by organisms in response to complex environments.

The reason Spencer will occupy an important place in this book is not because I think of him as a forgotten genius. I do not think this (although I do think he is a more interesting thinker than he is usually thought to have been). Spencer's work is important here for several reasons. First, his work was central to the initial meeting of two massive currents of thought. Though he was an evolutionist, Spencer's epistemology in general was continuous with the British empiricist tradition, stressing the role of experience and simple rules of association of ideas. Spencer stands exactly at the point where British empiricism met the theory of evolution.

His thought is also distinguished from earlier empiricist views with respect to several important features. First, Spencer was willing to theorize about the role played by the structure of agents' environments. Classical empiricists like Locke and Hume generally did not attempt to explain characteristics of thought in terms of specific characteristics of the external world. They start the story at the point where a sensory impression has appeared in the subject, and proceed from there. Spencer starts the story with the characteristics of the subject's environment that are perceived and dealt with.

Second, although Spencer was a broadly empiricist thinker he was not hostile at all to the idea that the mind has rich innate structure, as "rationalist" philosophers claim. Spencer was entirely prepared to concede this, as long as the innate structure has an adaptationist evolutionary explanation. Individual learning and adaptive evolution were, for Spencer, two specific modes of the same process. Recognizing adaptive innate structure was no concession to Spencer's real opposition, people who would explain life and mind in terms of internal drives and principles. Spencer recognized the deeper concord of explanatory pattern in empiricist and adaptationist explanations, and he fused these mechanisms to produce a single externalist theory of the sources of organic complexity.

Spencer's evolutionism led him also to focus on the role of cognition in coordinating behavioral responses to environmental conditions. Here Spencer's views bring him close to some contemporary naturalistic thinking about the mind. Since the advent of behaviorism and, more importantly, philosophical functionalism, materialist theories of mind have focused upon the role mental states play in mediating between sensory input and behavioral output (Putnam, 1960; Armstrong, 1968; Lewis, 1972; Fodor, 1981). Further, in the past decade or so a number of thinkers have argued that a specific appeal to evolution, and to a concept of function with an evolutionary backing, is the key to overcoming a range of problems materialists face, especially problems involving the semantic content of thought (Millikan, 1984; Papineau, 1987). So some of the distinctive ingredients of Spencer's view are also key features of modern naturalistic theories of mind.

Sometimes people associate an empiricist temperament, combined with a focus on behavior, with the term "pragmatist." This is an oversimplification. In fact, William James developed parts of his view of the mind in *reaction* to Spencer's more scientistic views. Spencer has a historical role in the development of pragmatism, but not because he was a pragmatist. It is better to regard Spencer as an evolutionary naturalist.

One person who was a pragmatist, however, is John Dewey. In fact it is common to regard Dewey's later thought as the high point of the pragmatist tradition so far. I agree with this assessment. Like his predecessors Peirce and James, Dewey understood beliefs as guides to action, as instruments. He saw inquiry as a goal-directed attempt to relieve the anxieties of doubt and obstructed action, and as a tool for overcoming practical problems. Unlike Peirce and James, however, Dewey explicitly described intelligent action as a response to problematic environmental *situations*. Dewey related the goal-directed operations of thought and action to problems deriving from the fact that agents inhabit environments which are, in themselves, precarious and uncertain in many respects. In other ways these environments are stable, contain useful constancies. Intelligent agents make use of the stable patterns in order to respond intelligently to problems posed by the world's instabilities. That is, Dewey's account of the function of thought focuses, like Spencer's, on the role played by cognition in responding to a complex or variable environment.

This focus on specific environmental patterns is a distinctive part of both Spencer's and Dewey's epistemological views. It is common to think that very *general* features of the external world are relevant to epistemology; it is common to say that if the everyday world is composed only of material particles with primary qualities, or is composed only of particular things rather than universals, or is a domain characterized by change, then this makes a difference to theories of knowledge. Global claims about our epistemic abilities are often buttressed with global claims about the composition of the world. This is different from finding specific and contingent facts about how the world's furniture happens to be arranged, and developing an epistemology around these.

Thus the idea that mind is for dealing with complex environments marks a point at which a certain evolutionary continuation of orthodox empiricism, exemplified by Spencer, makes contact with classical pragmatism, exemplified by Dewey. Spencer and Dewey do not understand the thesis in the same way, and in many respects their philosophies could hardly be more different. But their acceptance of this claim about the mind marks a point of contact between them, and distinguishes their views of the mind from other approaches, including other empiricist, naturalist and pragmatist approaches.

This brings us to the point where I can say why this book will single out Spencer's and Dewey's views for detailed treatment, rather than give a general account of the history of theories that link mind and environmental complexity. The environmental complexity thesis marks a point of contact between Dewey's pragmatist philosophy and a more overtly materialist, naturalistic approach which makes use of evolutionary concepts. In my view, naturalistic materialism and Dewey-style pragmatism are the two most important rival philosophical outlooks which exist at present, as far as core metaphysical and epistemological questions about the relations between mind, knowledge and reality are concerned. It is these two outlooks whose agreements and disagreements are the most important to understand.

This is not to say that Spencer's ideas will be used here to represent the views of modern evolutionary naturalism. A very large proportion of Spencer's specific views on the relations between mind and environmental complexity, and on evolutionary and psychological mechanisms, are false. Most contemporary naturalistic philosophers would also find his overall world view a completely unacceptable one. But a philosophical system does not have to be true to be worthy of study, and useful in the task of developing and presenting difficult ideas. Spencer's views do provide a good starting point in the investigation of the relations between mind and environmental complexity. They also furnish ideal examples of externalist patterns of explanation, and they are important in understanding certain aspects of Dewey's thought.

I said that Dewey's work is the high point of the pragmatist tradition, and that I aim to investigate the relation between naturalism and pragmatism. But Dewey often used the term "naturalistic" for his own epistemological view, and he was guarded with the term "pragmatist." Certainly Dewey was closer than other pragmatists to modern naturalistic views. However, there are definite differences between Dewey's views of mind and knowledge, and those of contemporary naturalists such as Dretske, Fodor, Millikan, Kitcher, Devitt and Goldman. It is this recent tradition, rather than "naturalism" in the abstract, that I am interested in. Contemporary naturalism stresses the continuity of philosophy with science (Kornblith, 1985; Kitcher, 1992). Philosophical questions are viewed as very abstract scientific questions, and are treated in these terms. But from a perspective such as Dewey's, modern naturalists are too conservative; they are too prepared to accept the categories and terms of debate which the philosophical tradition has bequeathed. They use science to attack the old problems of mind, representation and knowledge head-on, for the most part. For Dewey, these problems are set up by the tradition in such a way that there can be no solution, and many standard philosophical outlooks, including familiar forms of materialism, are designed to paper over the real problems, to make possible a systematic shirking of the concrete issues that philosophy should face. Thus Dewey's version of the environmental complexity thesis is intended as part of a more radical reorientation of philosophical theorizing about knowledge. One aim of this book is to use the environmental complexity thesis as an instrument for exploring where Dewey's pragmatism and recent naturalism converge and diverge, on questions about the relation between thought and the world.

So the central aims of this book are nonhistorical; they have to do with developing the resources of the environmental complexity thesis, understanding the logic and pattern of externalism, and investigating the relations between naturalism and pragmatism. But there will be a number of historical discussions. Historical examples will first be used to illustrate externalism and the characteristic patterns of debate between externalists and their opponents. Second, we will look at parts of the world views of Spencer and Dewey, and see the role played in these views by versions of the environmental complexity thesis. The historical discussions are not intended to give a complete history of ideas about the relation between mind and environmental complexity, and they are not intended to cover every aspect of Spencer's and Dewey's views of the world. The aim is to look at the development of a specific line of thought about the function of mind in nature, and then to improve this idea.

1.3 Outline of the book

The remainder of this chapter, and all of Chapter 2, are concerned with laying out the general framework used in the rest of the book. This first chapter will examine some of the key terms used in my statement of the environmental complexity thesis. In particular, it will outline how "complexity," "function" and "cognition" will be understood. Two versions of the thesis, the teleonomic version and the instrumental version, will be distinguished. Then Chapter 2 will look in detail at the externalist approach to organic systems, as seen across a variety of fields in the sciences and philosophy, and at the ongoing oppositions between externalist and internalist patterns of explanation.

The third chapter will examine Spencer, and the fourth will look at Dewey. These chapters will discuss two very different versions of the environmental complexity thesis, and describe how this thesis has a place in two different overall views of mind and nature. As we go through these chapters we will extract specific aspects of Spencer's and Dewey's versions of the thesis for positive endorsement. Dewey's position contributes more to this project than Spencer's.

With the basic outlines of a version of the environmental complexity thesis laid out, Chapters 5 and 6 will look at two specific concepts which may be useful in understanding the relations between organisms and environments: the concepts of *construction* and of *correspondence*.

Both concepts are very philosophically controversial. If any term can raise small hairs on the back of contemporary necks, it is "construction." Concepts of construction and constructivism have become new battlefields on which age-old conflicts between realists and their foes are now fought. These are also stamps on the side of large crates of new perspectives and methodologies which proliferate through the humanities and social sciences. Many who call themselves constructivists oppose the very idea of a "given" and natural world, which languages, belief systems and social practices conform to or reflect. Instead reality is constructed by these languages, belief systems and patterns of practice.

Dewey, too, thought that construction was a critically important relation between inquirers and their environments; thought functions to aid in the reconstruction of conditions, and the transformation of situations. If we accept

9

a Dewey-style version of the environmental complexity thesis, does that mean we also have to enthuse about the "construction" of environments? Yes, but only if construction is understood in the right way, and this is a way that does not have the ontological sting of contemporary constructivism.

If pragmatists and other malcontents emphasize construction, the more orthodox (including more orthodox naturalists) often hold out for *correspondence* as a relation between internal and external. A thought has the highest credentials if it corresponds to the world. Spencer was an enthusiast for relations of correspondence, and he gave this concept a central place in his theory of mind. Pragmatists universally distrust correspondence as a relation between internal and external, and Dewey was no exception. But both Dewey and Spencer lacked a moderate, modern understanding of what correspondence as a property of thought might be, informed by recent work in naturalistic philosophy of mind. Once we have such a concept of correspondence, should the environmental complexity thesis be understood as vindicating correspondence or replacing it? This is a difficult question, which I place at the end of Part I of the book. It also is a point at which the disagreement between modern naturalism and pragmatism is sharp. My tentative verdict is that correspondence can be retained by the naturalist. But there is also a possibility that it can only be retained at a cost. Once naturalized, correspondence may lose some of its apparent explanatory importance.

So the first part of this book is structured in pairs of chapters. There are two chapters which lay out the framework used, two chapters on Spencer and Dewey, and two chapters on specific concepts which might be important in understanding organism/environment relations, the concepts of construction and correspondence.

The second part of this book is focused on models. Some ideas which were discussed in Part I in an informal way will be recast in a simple mathematical form in these later chapters. Chapters 7 and 8 will outline the framework of a very general theory of adaptive response to environmental variability, a theory which includes intelligent, cognitive modes of response but includes flexibility in development and physiology as well. Chapter 7 makes use of recent biological models of "phenotypic plasticity," models intended to describe the circumstances under which a plastic or flexible organism has higher fitness in a variable environment than an inflexible one. Chapter 8 embeds these results within another mathematical framework, "signal detection theory." This chapter will look closely at *reliability* as a relationship between organism and environment. Reliability will be construed as something the organism can "invest" in, or not, depending on the relation between the costs and benefits of tracking the world.

One specific, simple form of question is central to these chapters. We will look at cases in which an organism is located in a complex environment, an environment which can be in a number of possible states. We will ask about the circumstances in which this environmental complexity should engender complexity within the organic system. When should heterogeneity in the environment bring about heterogeneity in the organism's response? When should the organism attend to the differences between environmental states, and when should it ignore these differences, and act as if the environment was always the same? When should environmental complexity bring it about that the organic system will *make a distinction*, will attend to a difference in the world?

The framework used in these chapters is decision-theoretic, in a broad sense. It makes use mainly of simple statistical concepts along with assumptions about actions and payoffs. This is one specific way to follow up and formalize some of the ideas discussed in earlier chapters. It is not the only way, however. These chapters should be understood as illustrating one possible direction of further investigation into the relations between internal and external complexity. It is not claimed that the decision-theoretic approach is the only way to proceed further on these questions.

In Chapter 9 we will switch from looking at complexity or flexibility as a property of individual organisms to looking at population-level analogs of these properties. We will look at ways in which environmental complexity can generate complex populations, rather than complex individuals. This will be done via an examination of some models developed in genetics and population biology during the 1950's and subsequent decades. This is, for the most part, not a discussion of the environmental complexity thesis per se, but a discussion of a more general explanatory pattern which the environmental complexity thesis exemplifies. Chapter 9 will also look at an ambitious and philosophically rich attempt to give a general model of individual-level and population-level responses to environmental complexity: Richard Levins' book *Evolution in Changing Environments*.

1.4 Thought and act

At this point we can begin the task of taking apart the environmental complexity thesis and examining its components. Three issues will be discussed in the remainder of this chapter.

First, we can divide the environmental complexity thesis into two separate claims. If a link is to be established between cognition and dealing with a complex environment, it is very likely that this link will go via behavior. So there is implicit in the environmental complexity thesis a claim about the relation between cognition and behavior, and also a claim about the relation between behavior and a certain kind of environment. In this section we will look at the relation between cognition and behavior.

The second issue that will be discussed is the fact that the environmental complexity thesis, as it was presented at the outset, is apparently a teleological claim of some sort. It makes a claim about the "function" of cognition. The suggestion is not just that dealing with environmental complexity is something cognition happens to do, but that this is what cognition is "for," that this activity has a special status among the range of things it can do. The meaning and force of such functional claims has been extensively debated in philosophy of science, and it is necessary to take some care outlining how claims about functions will be understood in this book.

Third, the last section of this chapter will discuss the concept of complexity.

I said the environmental complexity thesis can be understood as asserting a connection between thought and behavior, and also a connection between behavior and the world. This is the most straightforward way to understand the claim that cognition plays a role in adapting agents to complex environments. But the claim that thought has a certain kind of effectiveness in the world only via its links to behavior should not be taken to be completely trivial or *a priori*. At least, it is not trivial unless behavior is understood in an extremely broad way.

In certain fish, capacities of perception and perhaps also inference are used in directly regulating central aspects of development, as well as behavior. These fish determine whether they will develop as male or female via perception of their relative size within the population (Francis and Barlow, 1993). Perception and processing in the central nervous system affect the production of hormones which in turn determine sex. Here cognition aids the fish in dealing with its environment – perhaps in dealing with environmental complexity – without this going via a link to behavior. Similarly, suppose it is true that cognition has, aside from any links to bodily motion, the effect of fine-tuning the hormonal balance in the body. Patterns of thought which induce certain moods bring about a better functioning of overall metabolism. If this were so, then at least part of the adaptive role of thought would be independent of behavior (and there would be some slight vindication of those Ancients who thought of the brain as a sort of radiator).

Naturalism and teleology

So the interpretation of the environmental complexity thesis which makes an essential appeal to behavior is not strictly speaking the only possibility. In this book we will focus on cognition as a determinant of behavior, however. This decision reflects a bet about what cognition is *mostly* concerned with, not the acceptance of an *a priori* principle about the necessary role of cognition. Though that is the official position of this book, much of the discussion of the role of thought and its impact on the world does not make specific assumptions about what exactly constitutes behavior. If behavior is understood in a loose way (even if it is taken to include the developmental choices of fish) this will not affect most of the claims made below.

Most of the discussions of "cognition" and "thought" in this book will be concerned with a small set of the range of properties and processes normally considered "mental." We are concerned with a basic cognitive tool-kit enabling perception, the formation of beliefs or belief-like representational states, their interaction with motivational states, and behavior. While we will discuss thought in the context of its relation to behavior, the nature of this relation will be different from that seen in many other recent discussions of the mind.

Most recent philosophy of mind has been "functionalist" in some sense or other. We can distinguish two basic forms of functionalism in philosophy of mind. First there is the more orthodox view which I will call "dry functionalism." This view understands function in terms of causal role, and it identifies mental states in terms of their typical causal relations to sensory inputs, other mental states, and behavioral outputs (Block, 1978; Lewis, 1980). Second there is "teleo-functionalism." This view makes use of a richer, biological concept of function more closely allied to traditional teleological notions, a concept often analyzed with the aid of evolutionary history (Lycan, 1981; Millikan, 1984; Papineau, 1987). For the dry functionalist, one essential property of any mental state is the pattern of behavioral outputs which the state, in conjunction with the rest of the system, tends to cause in various circumstances. For the teleo-functionalist, what is essential to the mental state is not what it *tends* to do but what it is *supposed* to do.

For both dry functionalism and teleo-functionalism, links between mental states and behavior are part of what determine the identity of the mental state; they are part of what makes it the particular belief, fear or plan that it is. In this book there will generally be no attempt to forge these constitutive links between thought and behavior. The environmental complexity thesis is not supposed to solve the problem of locating the mental in the physical world. The aim is rather to discuss the *role* of thought in nature, and the existence of the two will generally be taken as given.

13

Of course, it is part of the point of functionalism, both dry and teleological, to say that specifying something's role *is* specifying what it is, for certain sorts of entities. For a functionalist, the question of how the mental exists in the physical and the question of what the mental does are in large part the same question. But though I have sympathy for this project, this book is not generally concerned with a functionalist project of that sort.

Similarly, the term "pragmatism" is sometimes used for the view that beliefs are to be individuated in terms of the habits of behavior they produce (Peirce, 1877) or in terms of the conditions under which the behavior produced is successful (Ramsey, 1927). This book will not, in general, understand pragmatism to be a theory of belief content. The only part of the book which will examine the role of behavior in fixing meaning is Chapter 6, which discusses the concept of correspondence. There it will be necessary to look closely at behavior-linked theories of belief content. But in the rest of this book we will not be concerned with constitutive links between thought and behavior.

So we will understand the environmental complexity thesis as asserting a connection between thought and behavior, and a connection between behavior and environmental complexity, although neither connection needs to be understood as constitutive of any of the entities involved. Now we need to look closely at what sort of connection this is, and at the term "function" in my statement of the thesis.

1.5 Two concepts of function

We are discussing a possible view about the "place of mind in nature," where finding mind's "place" is, again, not understood as finding where it exists in the physical world, but finding its characteristic *role* in nature. Let us suppose, as is plausible, that there is a long list of things cognition actually does in the world. It can have a different role in different sorts of people, different sorts of animals, in the same person at different stages in life, and so on. In the face of this diversity, can there be a single, abstract philosophical claim made about "the" role of cognition?

It is very possible that the answer is no, but if that is so the answer will be found by trying and failing. It is not true that attempts to single out some distinctive role or function from a multiplicity of effects and dispositions are always fruitless. This problem is constantly encountered by biologists of various kinds, and in the course of inquiry in biology (and various other sciences) principled ways have been developed for distinguishing the function of a structure or trait from mere effects. Philosophers of science have become

14

intrigued by this problem, and by the persistence of apparently teleological language in fields such as evolutionary biology. This philosophical attention has isolated two distinct concepts of function, which are acceptable to the naturalist and which can be used here in analyzing the environmental complexity thesis. The critical thing is to recognize the distinctness of these two concepts, and the different roles they play.

The two analyses of function which will be used here are basically those of Larry Wright (1973) and Robert Cummins (1975). Although it is common in the philosophical literature to understand these two views as competing analyses of a single concept, there is no obstacle to accepting both concepts as legitimate. On this view, "function" in scientific contexts is ambiguous – or rather, more ambiguous than it is usually thought to be.

The two concepts can be presented in a way which shows their compatibility by focusing on the problem of function as it was presented above, as a problem of selecting the function of a trait or structure out from a multiplicity of effects. To use an example which has long played the role of a test case, how is the function of the heart picked out from its long list of effects? Why is the heart's function pumping blood, when it also has the effect of making certain sounds, of filling a certain space in the body, and so on? Wright, in 1973, proposed that functions are distinguished from mere effects by their explanatory significance. The function of something is the effect it has which explains why it is there. Hearts are found where they are because they pump blood, not because they make certain sounds.

In 1975 Cummins argued that Wright, among others, had mistaken the type of explanation in which ascriptions of function are used, and he proposed an alternative analysis. Cummins' concept of function is not used in explanations of why the functionally characterized thing exists, but explanations of how some larger system, which the functionally characterized thing is part of, manages to exhibit some more complex capacity or disposition. So for Cummins, we single out the heart's blood-pumping function as part of an explanation of how the body as a whole achieves circulation of oxygen and other substances, not as part of an explanation of why people have hearts.

There is no need to choose between these two concepts. Functions, for both Wright and Cummins, are effects distinguished by their explanatory salience; this is the core common to the two concepts. Both the form of explanation discussed by Wright and that discussed by Cummins do exist in science; there are explanations of the presence of things in terms of what they do, and there are explanations of how systems realize complex capacities via the capacities of their parts. In the field of behavioral ecology, for instance, function ascriptions tend to have the force Wright ascribes (Allen and Bekoff, forthcoming). Functional characterization cites the effect(s) of a behavior pattern which explain its survival or maintenance under a regime of natural selection. To say that the function of a male lizard's display behavior is to intimidate other males is to say why the behavior is produced by the lizards. But in a field like neurophysiology, which is concerned with the behavior of a very complex system of interacting components, an ascription of a function often is used to single out an effect a component has which is salient in the explanation of some complex capacity of the system. Myelin is said to have the function of insulating nerve cells in such a way as to make possible long-distance conduction of signals. This need not be understood as a claim about why the myelin is there (though it could be). The claim has a legitimate place in a Cummins-style analysis of how the brain does what it does.

The analysis originally given by Wright had various problems of detail, which were discussed in subsequent literature.[1] His view is basically sound, however. The only fine-tuning I will make use of here is to explicitly require that for an effect to be a function, it must explain why the functionally characterized entity exists, and this explanation must involve some process of selection. This need not be natural selection on genetic variants, as in the paradigm biological cases. Conscious selection by a planning agent, selection through reinforced learning and other types of individual-level adaptation, and cultural processes of selection are also counted.[2]

This first concept of function will be called the *teleonomic* sense of function (Pittendrigh, 1958; Williams, 1966). "Teleonomic" is a term sometimes used by biologists to refer, roughly, to those parts of traditional teleological thinking that can be given a foundation in the operation of natural selection.

The analysis Cummins gave, of the other concept of function, can also be adopted almost intact. The only change made will be to drop the requirement that the capacity explained by a function is more complex than the function itself. So functions in this second sense are capacities or effects of components of systems, which are salient in the explanation of capacities of the larger system. These will be called "Cummins functions." An important feature of the concept of a Cummins function, which Cummins himself noted, is its undemanding nature. Cummins functions can exist wherever there is a system and a project of explanation directed upon the capacities of the system. Once certain capacities of the system are the focus of explanatory interest, and capacities of the components are relevant to explaining the capacities of the larger system, these capacities of the components become Cummins functions, relative to that explanatory project. The existence of a "system" can be a matter of choice or convention, as can the capacities of the system deemed worthy of explanation. The part which is not a matter of choice or convention is the real contribution the components make, or do not make, to the capacities of the system.

So a single entity could have a great many Cummins' functions – as many as there are systems it could be considered part of, and whose capacities it makes a real causal contribution to. I said earlier that the concept of function provides a means for distinguishing some out of the multiplicity of effects and dispositions which a thing might have. In the case of Cummins functions, the concept of function provides this distinction between function and effect in a way which is relativized to an explanatory enterprise and the specification of a larger system. As Cummins said, if our explanatory interest is directed upon the total noise made by a living human body, then the heart would have the (Cummins) function of making the thumping sounds. Teleonomic functions are not like this. They are determined solely by the explanation for the existence of the functionally characterized entity. No amount of deeming the human body to be a sound system can change the fact that hearts are not present because of their sounds. Teleonomic functions are far thinner on the ground than Cummins' functions, and their existence is a more objective matter.

Though the Cummins concept of function is accepted as it stands, a certain type of Cummins function is intended to be central in this discussion. These are cases which have an *instrumental* character. In these cases a Cummins function is a contribution made to a capacity or property of a system which is valued for some reason, often a property valued by the system itself.

These instrumental cases are ones in which Cummins functions capture many of the paradigm phenomena for "goal theories" of functions, theories holding that functions are contributions to a system's goals (Wimsatt, 1972; Boorse, 1976). Goal theories themselves are, among their other sins, too restrictive. Cummins-style functional analysis need not ascribe goals. The Cummins concept of function captures a type of analysis in which a theorist investigates the role played by a component in producing the distinctive properties of a larger system. Instrumental cases are one important type of case, but not the only type.

If functions are to rest upon explanation, it makes a difference how explanation is understood. For the purposes of this book, I will generally assume a causal view of explanation, according to which an explanation is a factor causally relevant to the event explained (Salmon, 1989). Among the factors causally relevant to an event, only some count as explanations in any given theoretical context. But *only* real causal factors are explanatory, and a realist view of causation is also assumed. This view may not capture every type of genuine explanation, but it is adequate to the cases we will be concerned with. Occasionally more specific assumptions will be made about explanation along the way.

On the view I have outlined, the only thing the two concepts of function share is that in both cases functions are explanatorily salient effects of components of systems. No doubt many will feel a lingering attraction to a more unified analysis.[3] But the disunity which the present concept of function has is not the product of philosophical difficulty, but a recognition of real differences in the explanatory projects which exist in fields such as biology.

These two concepts of function will form part of the framework within which the environmental complexity thesis will be investigated. These are the two senses in which the thesis will be understood as a theory of the "place" of mind in nature, or the "role" of cognition in the world. On the teleonomic reading of the environmental complexity thesis, the thesis is intended as part of an explanation of why mind exists in nature. The thesis cites one of the things cognition does, and claims that this disposition or effect is important in explaining why cognition exists. In this discussion I will generally leave it open whether the environmental complexity thesis is intended to explain how cognition originated, or why it has been maintained in existence, or both. This distinction is important, but it will not often affect the discussion here.

In the context of the second sense of function, the Cummins sense, the environmental complexity thesis is understood as a claim not about why cognition exists, but rather as part of an account of the role played by cognition in explaining some features of larger systems. The "larger system" assumed to be most important here is the whole organism which houses the mind in question. For example, in this second sense the environmental complexity thesis can be understood as part of an explanation of how organisms like ourselves succeed in getting by, in staying alive and achieving their aims. Note that while teleonomic functions refer necessarily to actual history, albeit very recent history in many cases, Cummins functions are independent of time.

I should try to make the motivation for adopting these particular concepts of function more intuitive. The idea is that there are two distinct interpretations of the environmental complexity thesis that are worth investigating. It would be valuable to know what cognition is *for*, and it would also be valuable to know what cognition is *good for*. As Wright argued, these are two distinct types of question which must not be conflated. Wright also argued persuasively that only his "etiological" analysis could capture the sense of function in which a function is what something is for. But once we know what cognition is for, why it is there, what is its *raison d'etre*, we can also ask what else cognition is good for, what other contributions it makes, what other uses it can be put to. There is no reason why something's *raison d'etre* should exhaust or constrain its instrumental role, its distinctive effects and uses once it exists. A car's heater is *for* keeping the passengers warm; but it can also be *good for* helping to keep the engine cool. Thus we recognize an instrumental version of the environmental complexity thesis as well.

A function in the instrumental sense, just as in the teleonomic sense, must be an effect that the entity in question actually has; it is not enough for the entity to be merely believed to have that capacity or effect. Neither is it enough for the entity to have the potential to have a certain effect, in hypothetical circumstances. So a possible but nonactual use of mind is not itself a function. However, an investigation of the actual functions of cognition, the actual role it plays in the world, has the potential to unearth new applications and uses as well.

The teleonomic sense of function, and the Cummins sense, will provide the *only* senses in which the environmental complexity thesis will be interpreted, as far as the positive project of this book goes. These are the only senses in which we will aim at a theory of the "role" of mind in the natural world. What distinguishes a function from a mere effect is causal/explanatory importance and no more.

This point is important, as it is easy to allow the naturalistic concept of function to grow illicitly into something larger. A distinctive part of the biological concept of function is that where there are functions there can be *mal*functions. The concept is normative in this way (Millikan, 1984). So if we have a theory of the function of cognition we have a theory not just of what cognition actually does, but also of what it is *supposed* to do, what activities are "proper" to it. This is right, but we must pay constant attention to the exact sense in which naturalistic concepts of function license normative language. In the teleonomic sense of function, for something to perform its function is for it to do whatever explains why it is there. To be in the relevant circumstances but do something different is to malfunction.[4] In the Cummins sense, for something to malfunction is for it, when in the relevant circumstances, to do something other than the thing which explains how this component contributes to some particular capacity of a larger system. In both cases, to malfunction is simply to *fail to do the explanatory thing*. This is a very weak kind of normativity.

Some might wonder, at this point, why failing to do the explanatory thing should even be called "malfunction." That term suggests failing to do the right or appropriate thing. Why should the explanatory thing be the right thing? There is, in general, no reason why it should be. The only circumstance where ascriptions of function and malfunction have real evaluative force is a special case, the case where part of the context is constituted by the interests of an agent, or where for some other independent reason, the effects which are functions are valuable ones. Teleonomic functions exist as a consequence of processes of selection. One important kind of selection is the kind where an agent has desires and goals and makes a selection in order to pursue those desires. In the case of a museum guard, doing the thing that explains why he is there will typically coincide with doing the right or appropriate thing according to the interests of his employers. Similarly, in the context of the Cummins sense of function, if the capacity of the larger system in question is something valued by an agent, such as the security of the museum, then when a component neglects to make its characteristic contribution to the realization of this capacity, failing to do the explanatory thing is failing to do the right thing, by someone's standards. These are important special cases, and the instrumental cases of Cummins functions tend to fall within this class. But there are other cases as well, in which the processes of selection are natural rather than conscious, and in which a system's capacities are not valued by anyone. In these cases "malfunction" means "failure to do the explanatory thing" and no more.

There are some complex problems of detail concerning the relation between attributions of function and the intentions of agents (Wright, 1976). Some common-sense intuitions about teleological concepts may well form persistent obstacles to all naturalistic analysis. In this discussion, the most important thing is to isolate senses of "function" which are clear, naturalistically acceptable, and which can be used in developing the environmental complexity thesis. This is what the teleonomic and instrumental senses of function are intended to be. As it happens, some parts of the discussion which follows are insensitive to substitutions of other concepts of function. Much of the time we will simply be looking at some distinctive things which cognition does, and the specific relation between effect and function will not often be an issue. However, it is important for anyone interested in devising a naturalistic view of mind's function or role in nature to have a position on what such a theory is supposed to be saying.

In the next section we will look in more detail at how functional claims about the relation between cognition and behavior are understood within the analysis I have outlined.

1.6 Teleonomic and instrumental views of cognition

Here is a well-known passage from William James, which appears in his essay "The Sentiment of Rationality":

> Cognition, in short, is incomplete until discharged in act; and although it is true that the later mental development, which attains its maximum through the vast hypertrophied cerebrum of man, gives birth to a vast amount of theoretical activity over and above that which is immediately ministerial to practice, yet the earlier claim is only postponed, not effaced, and the active nature asserts its right to the end. (1897b, p. 85)

Suppose we are considering whether to endorse this claim. In the framework laid out above, there are two legitimate ways to understand this passage. Neither is likely to be exactly what James intended, but these are the only two ways recognized in this book. We can first understand the passage as making a claim about the teleonomic function of cognition, its *raison d'etre*, the thing it does that explains why it is there. The claim is that cognition has the function of guiding behavior, and that any particular instance of cognition has not performed its function until it has made some difference to behavior. Read in this way, the passage makes a straightforwardly empirical hypothesis about why thought exists.

Second, we can understand James' claim with the aid of the Cummins conception of function. Now we are interested not in why cognition exists, but in what it is doing here, what sort of contribution it makes to the properties of larger containing systems. If read this way, the claim in the passage is also very strong. It says that cognition makes no distinctive contribution to anything, except insofar as it influences behavior. Aside from its effects on behavior, cognition is an idle part of the world.

In order to make the relation between the two readings clear, consider analogies between thought and other human activities such as eating and having sex. Speaking teleonomically, sexual intercourse is incomplete until discharged in fertilization, and eating is incomplete unless the body is recharged through nutrition. These seem obvious teleonomic claims – if they are not obvious then grant them for the moment. Against this basic teleonomic background, there will be additional teleonomic facts explaining details of the gustatory and sexual acts which people engage in, having to do with social bonding, ritual, the maintenance of power relations, and so on. That does not alter the basic teleonomic facts about these activities, the reasons why they are there at all. It is also true that the later developments of these activities, which

21

attain their maximum in the hypertrophied practices of humans, giving rise to a vast amount of activity not immediately ministerial to nutrition and reproduction, do not efface or negate the basic teleonomic claims. But these teleonomic facts do not in any way exhaust the *instrumental* roles which food and sex can have. They can be used for other purposes, can contribute to other systems, can have distinctive effects in a great variety of other domains.

Once we have separated the two types of functional claim sharply, it is apparent that the environmental complexity thesis is a very strong claim, on both the teleonomic and instrumental construals.

Interpreted teleonomically, the environmental complexity thesis is a historical claim of a particular sort. It is an explanation of cognition's existence in terms of certain of its effects. On this interpretation the thesis has a great deal of empirical content, implying claims not just about what cognition has done in the past, but also about the circumstances surrounding this history. The development of thought out of the unthinking, and its maintenance, have presumably been the consequence of evolution by natural selection. When an evolutionary explanation is given which simply mentions certain alleged benefits associated with a trait, rather than a detailed array of evolutionary forces, constraints and initial conditions, the explanation is often referred to as "adaptationist." I am understanding one version of the environmental complexity thesis as an adaptationist hypothesis.

While adaptationist thinking is associated with a range of famous pitfalls (Gould and Lewontin, 1978), one problem is especially relevant here. This concerns the individuation of the trait under discussion. As it was initially formulated, the environmental complexity thesis refers simply to "cognition," as if this was a single thing, a single trait. However, the generalized category of "cognition," even if it picks out a single kind relevant to everyday discussion, may not reflect a single evolutionary reality. "Cognition" may well be a collection of disparate capacities and traits, each with a different evolutionary history.

If we look at the totality of mental life, this seems a likely outcome; it would be surprising if there was a single evolutionary story that accounted for perception, consciousness, planning, mental representation, pain, dreams, emotion, and so on. As outlined earlier, many of these aspects of the mind will not be discussed in this book; we will focus on a small and allegedly fundamental set of mental phenomena. We are concerned with a basic apparatus that makes possible perception, the formation of belief-like states, the interaction of these states with motivational states such as needs and desires, and the production of behavior. Even this contracted set may not constitute anything like a single trait, for evolutionary purposes, but we will proceed on the assumption that it does.

Some might take this restriction to bias the case in favor of the environmental complexity thesis at the outset; we have restricted ourselves to just those aspects of mental life that *do* seem to have some connection with behavioral problem-solving. We have excluded those aspects – like the qualitative feel of colors – that do not. This accusation is partly true; the aspects of mentality we will focus on are the ones that apparently lend themselves to an analysis in terms of environmental complexity. The selection of this subset of mental capacities is not arbitrary, however. On the environmental complexity thesis, these aspects of mentality are hypothesized to be the teleonomically fundamental ones. Other properties of thought are viewed as teleonomically secondary; the *primary* teleonomic fact about cognition is that it is a means to the production of behavioral complexity, and this enables agents to deal with environmental complexity. The fact that certain other properties of the mind may be teleonomically "secondary" does not mean that they are secondary in any other sense.

It will also be assumed that this basic cognitive apparatus seen in people is continuous with the apparatus displayed in less complex animals, including not just familiar vertebrates like cats and monkeys but also creatures with far simpler nervous systems. The environmental complexity thesis will be understood, in its teleonomic sense, as an adaptationist hypothesis about a core set of capacities that nervous systems make possible and which are displayed by a large range of animals.

This is certainly a large set of assumptions to make. I am not trying to damn the principle at the outset, though. "Camouflage" and "swimming" can be regarded as single, very widespread traits, which can be realized in diverse structural properties, and about which general but plausible adaptationist arguments can be made. It is not hopeless to develop a general theory of the place of swimming in nature, even in the absence of detailed knowledge of genetics and the historical relations between different lineages of animals. There is a single class of environments in which the capacity for swimming is favored – liquid or partly liquid environments – and despite the details of its realization in different animals, there is also a set of teleonomically fundamental properties of the capacity for swimming. Cognition might be similar; it might be a single trait with certain teleonomically central features, and with a small set of evolutionary factors bearing consistently upon it across different lineages. On the other hand, it might be a conglomerate of traits, located in a tangled nexus of evolutionary forces, different in every case, and defying

23

simple-minded adaptationist analysis. This book does not aim to decide which of these is true. It will explore a set of broadly adaptationist ideas about the mind, treating a basic cognitive tool-kit as a unit for adaptationist analysis. This assumption about the teleonomic unity of the tool-kit is not regarded as obvious or necessarily true, but as an assumption which opens up a field of possibilities worthy of philosophical exploration.

Lastly, it should be stressed that the environmental complexity thesis in this sense is construed as an adaptationist claim about the capacity for cognition itself; it is not itself an explanation for particular thoughts or patterns of belief. How much of the day-to-day operation of the mind can be explained in an adaptationist way is a further issue. If the environmental complexity thesis is true, then somehow cognition must help agents to deal with environmental complexity. But *how* this is achieved is a different question.

1.7 A simple concept of complexity

The thesis we are investigating says that cognition has a certain relation to environmental complexity. Cognition, too, is a complex thing. So one way to look at the environmental complexity thesis is as a claim about a relationship between internal and external complexity. Part of this book will pursue this line of thought; the thesis will be seen as exemplifying a more general view about organic complexity and its relation to environments.

To do this we must look first at the concept of complexity. In this book complexity will be understood as *heterogeneity*. Complexity is changeability, variability; having a lot of different states or modes, or doing a lot of different things. Something is simple when it is all the same.[5]

In this sense, complexity is not the same thing as order, and is in fact opposed to order. Heterogeneity is disorder, in the sense of uncertainty. The concept of complexity as heterogeneity is a scientifically useful one, as this property can be measured. One way to measure it is with the "Shannon entropy" of information theory.[6] More simply, in J. T. Bonner's book *The Evolution of Complexity* (1988), complexity of form is measured as the number of different cell types in an organism. In different circumstances different exact measures of complexity can be relevant, but the core of the concept is heterogeneity.

If complexity is understood as heterogeneity or variability, then both an organism and an environment can be said to be complex, or simple, in the same sense. An environment with a large number of different possible states which come and go over time is a complex environment. So is an environment which is a patchwork of different conditions across space. The heterogeneity property is not the same in these two cases, but in both cases heterogeneity can be opposed to homogeneity. A complex environment is in different states at different times, rather than the same state all the time; a complex environment is different in different places, rather than the same all over. Similarly, a complex organism is one which has a heterogeneous structure or is heterogeneous with respect to what it does. A simple organism is homogeneous – uniform in how it is put together, or in what it does.

To say that complexity in an organism is understood as heterogeneity is probably to leave out some aspects of the common sense image of organic complexity, if there is such a thing. When attributed to an organic structure, complexity in common sense discourse probably involves not just heterogeneity but also some property of organization. Mere disorder in an organism might not normally be recognized as biological complexity. Perhaps the common sense view is that complexity is *organized heterogeneity*. If organization is understood as a property of regularity or predictability, then biological complexity would be some combination of order and disorder, homogeneity and heterogeneity. It would be a major task to make this concept completely clear. In this book we will abstract away from the property of organization as much as possible. The focus will be on the aspects of complexity that can be understood, and measured, as heterogeneity or variability alone.

Although whenever complexity is discussed in this book it will be understood as some property of heterogeneity, it is not assumed that there is a single measure of complexity which applies in every case. There are many different types of heterogeneity, both with respect to properties of organic systems and with respect to environments. Any environment will be complex in some respects but simple in others, and only some complexity properties will be relevant to any given organism. An environment may be simple at one level or time-scale but complex at another, for example. The status of complexity properties will be discussed in more detail in Chapter 5.

Lastly, it will be useful to make at this point two basic distinctions between different types of organic complexity. First, we can make a distinction between *first-order* and *higher-order* properties of complexity, variability or flexibility.

Complex behavior is heterogeneous behavior, doing different things in different circumstances. But consider two different types of behaviorally variable organism. One is "smart" in the sense that it can track the state of the world, and react to changes in its environment with appropriate behavioral adjustments. But the set of rules or conditionals – "If the world is in S_1, then do B_1" – which determine which behavior is produced in each situation, is fixed.

This organism is behaviorally complex when compared to an organism which does the same thing in every situation, which performs the same action come what may. The organism which adjusts its behavior to circumstances, but does so in a rigidly preprogrammed way, has a *first-order* property of complexity in its behavior.

Such an organism is inflexible in contrast to an organism which is able to modify its behavioral *profile* in the light of experience, an organism which modifies what behavior it is that is produced in the presence of a given environmental condition. This second type of organism is able to change the list of conditionals that determine what it does in a given situation. This is learning; the learning organism can learn that it is not so good to produce B_1 when the world is in S_1, and better to produce B_2 instead. This is a *second-order* property of complexity.[7] There is also third-order plasticity, the ability to change the learning rules which are used to determine the list of conditionals ... and so on.

When complexity of behavior is measured as sheer variety, a merely first-order complex organism may produce more complexity than a second-order one. A first-order complex organism might have dozens of behaviors available, while a second-order one might have only a few.

Second, we should distinguish between *structural* and *functional* complexity. Structural complexity has to do with what a system is made up of; functional complexity has to do with what it does, or is supposed to do.[8] Structural complexity has to do with both how many different parts there are in a system and also how these parts are put together, how they interact. Is it the sheer number of parts, or the number of types of parts? The latter is more important, though the former is also relevant in some contexts.

This book is concerned almost solely with functional complexity. In particular, we will focus on the capacity of certain organic systems to produce a diverse range of behaviors. Cognition will be viewed as a *means to the production of behavioral complexity*. Behavioral complexity is a particular type of functional complexity; it is the ability to do a lot of different things, in different conditions. It is not claimed here that intelligence and biological complexity, on some independent measure, are firmly correlated. There is also no attempt to regard either complexity or intelligence as a measure of evolutionary progress.

In part these ideas are rejected because of the difficulty involved in formulating a single measure of biological complexity; complexity is heterogeneity, but there are lots of heterogeneity properties. Similarly, it is not assumed that there can be a single useful measure of intelligence or cognitive capacity; there are lots of intelligence properties as well. But it is assumed here that cognition is a particular complex set of organic capacities, which make behavioral complexity of various kinds possible. We will ask about the properties of environments that tend to make this form of functional complexity worthwhile, and about the philosophical consequences of these connections between cognition, behavior and environment.

It might be thought that, in general, being able to do a lot of different things requires having a complicated mechanism with lots of parts – that functional complexity requires structural complexity. There are many cases in which this principle holds, but it must be treated with caution. Sometimes flexibility in behavior is a consequence of structural simplicity. For example, in an earlier section of this chapter I discussed the fact that some fish use perception and information-processing in sex determination. There are also fish which can *change* their sex individually within their lifetime, in some cases on quite a rapid time-scale (Francis, 1992). Individuals here are plastic with respect to their sex. On my account, this is functional complexity the fish have which all mammals, individually, do not have.[9] At the level of the *population* there is no difference in complexity between a fish population of this sort, and a mammal population such as the human population. Both populations have two biological sexes. But in the case of the fish, the population's complexity is a consequence of the complexity of the individuals. In the case of humans, the individuals are simple in this respect and the population's complexity is irreducible.

The case is relevant here because it may be that one reason the fish have this individual complexity is the fact that their developmental program and the relevant parts of their physiology are very *simple*, compared to ours. Our sexual development and physiology are so complicated that it would be very difficult to switch back and forth. In certain mammals, such as chimps, sex-change as a strategy might in fact do very well under natural selection (Francis, in preparation), but this may not be a developmental option. If so, this is a case in which structural simplicity engenders or makes possible a property of functional complexity, in the fish, and structural complexity engenders functional simplicity, in mammals. The two are not always correlated.

In the domain of behavior and nervous systems there may be more truth to a general principle that structural complexity is required for functional complexity. For example, some lines of research into connectionist models of cognition, in which thought is modeled in terms of interactions between idealized neuron-like "units," have produced definite principles linking the complexity of the structure of networks of units to the complexity of the problems they can solve (see Rumelhart et al., 1986, Chapter 8). These issues about structure are important, but in this book we will be concerned mainly with functional properties rather than structures and mechanisms.[10] The aim is to look at some fundamental properties of the relations between internal and external complexity, to treat cognition as a means to behavioral complexity, and to try in this way to cast light on the relations between mind and the rest of nature.

Notes

1 For discussion and elaboration of Wright's analysis see Boorse (1976), Millikan (1989a), Neander (1991), Griffiths (1992a) and Godfrey-Smith (1994a).

2 Wright, in his original discussions (1973, 1976), seemed to envisage that the only processes by which the effects of a structure would lead to its maintenance, its production, or its coming to be where it is, will involve the operation of some form of selection. But a thing can have effects which *prevent* the operation of a selection mechanism, and these effects can explain the thing's maintenance without being genuine functions. This is how some of Boorse's (1976) counterexamples to Wright's analysis work.

3 The best unified analysis I know is that of Kitcher (1993). If some would prefer to swap an analysis of functions like Kitcher's for mine, some (but not all) of the argument in the rest of this book can go through as before.

4 There are some issues of detail I am neglecting here, which do not affect the argument. If the structure in question does the usual thing but the environment does not cooperate, this is not a malfunction of course. Also, something might be selected because it does something useful very occasionally, in a random way, and does something pointless but harmless the rest of the time. This idle but harmless behavior is not malfunctional. In this case a probabilistic analysis is needed; the structure malfunctions when its probability of producing the useful behavior becomes markedly lower than usual.

5 See McShea (1991) for a useful review of concepts of complexity in biology.

6 For a good philosophical discussion of information theory, see Dretske (1981). The Shannon entropy is defined as follows. If there is a range of alternative states which a system can be in, and the probability of its being in state i is P_i, then the complexity or disorder of the system is measured as: $E = -\sum P_i \log_2(P_i)$. This measure has a high value when there are lots of alternative states that have similar probabilities, low when there are few states or some small number of states exhaust most of the probability. So this measure describes the unpredictability of the state of the system.

7 Mayr (1974) would describe first-order complexity with respect to behavior as a "closed program" and higher-order complexity as an "open program." The present

sense of "first-order" complexity has no relation to the "first-order" property of a first-order Markov process, as described in the theory of stochastic processes.

8 This is a point where the distinction between teleonomic functions and Cummins functions matters. Only variability in dispositions or effects which has a selective rationale is recognized as teleonomic functional complexity; Cummins functional complexity is more ubiquitous.

9 By "sex change" I mean here "functional sex change," which involves the capacity to engage in reproduction, not just change in structure, physiology and behavior. The same qualification applies to my claim that the human population contains only two sexes.

10 There are also conceptual questions about functional complexity even when we have set structural complexity to one side. A very simple program – like an ant's behavior program – can deliver complex output, if the input is complex in the right way (Simon, 1981). Is the ant functionally complex or not? It depends on which criterion of measurement is relevant, on what counts as "doing a different thing." Some of these issues about relevant differences will be discussed in detail in chapter 5. In Simon's case, I would say *to the extent that* the ant's behavior is recognized as complex (which depends on the criteria chosen) it is a consequence of environmental complexity. There is no complexity in the behavior which cannot be linked to environmental complexity.

2

Externalism and Internalism

2.1 Some basic explanatory forms

The environmental complexity thesis about cognition asserts a link between certain capacities of organic systems and a specific feature of these systems' environments. This is a familiar pattern of explanation. It exemplifies one of the basic stances or approaches that can be taken in investigating organic properties. The internal is understood in terms of its relation to the external. If a causal explanation is involved, the channel of causal influence goes "outside-in."

One of the aims of this book is getting a better understanding of this general pattern of explanation, and understanding some versions of the environmental complexity thesis as instances of this pattern.

Most generally, the term "externalist" will be used for all explanations of properties of organic systems in terms of properties of their environments. Explanations of one set of organic properties in terms of other internal or intrinsic properties of the organic system will be called "internalist."

These terms "externalist" and "internalist" apply to our explanations of properties of organic systems, not to the organic systems themselves. But an externalist claim can be true or false, and if a system, or some aspect of it, really is controlled entirely or for the most part by events in its environment, then an externalist pattern of explanation *applies* to it.

This chapter will mainly be concerned with the relations between internalism and externalism. A third basic pattern of explanation will be discussed in detail in Chapter 5, but I will introduce it now. "Constructive" explanations are explanations of environmental properties in terms of properties of an organic system. So this is the converse of an externalist explanation.

Fourth, within the general category of externalist explanations there is a subcategory we will look at in particular detail. These are explanations of properties of internal complexity in terms of environmental complexity; externalist explanations in which both *explanandum* (that which is explained) and *explanans* (that which does the explaining) are properties of complexity. These will be called "c-externalist" explanations.

Not all externalist explanations of complexity are c-externalist; it has to be an explanation in terms of environmental complexity. This "c-" prefix can be used generally for explanations of one type of complexity in terms of another. There are c-internalist explanations as well, and so on.

There are some explanatory relations which have not been named in this taxonomy. We have no term for an explanation of one set of environmental properties in terms of another set of environmental properties, the environmental counterpart of an internalist explanation, for example. But there is enough to keep track of here already, and these four categories of explanation – externalist, internalist, constructive and c-externalist – will be the ones we will be specifically concerned with.

So far these explanations have been described as if they were all exclusive of each other, as if believing that organic complexity is explained in terms of environmental complexity precluded one from believing that organisms affect the environment as well. We have not explicitly allowed for the possibility of two-way connections, such as feedback connections. First things first. In this chapter we will be concerned with internalist and externalist explanations considered individually. In Chapter 5 we will look at combination views, including the important combination where there is a two-way connection or "coupling" between organic system and environment.

2.2 A fast tour

There are few more fundamental features of intellectual life than the opposition between externalist explanatory programs and their internalist rivals.[1]

The opposition displays a similar structure across a range of sciences and also philosophy. It characterizes debates about the biological world, the mind, knowledge and also the social realm. The distinctive role the opposition plays is sometimes obscured by other perennial oppositions and debates, however. This is not the debate between reductionism and holism, or that between the friends of reason and science, and their enemies. Neither is it straightforwardly related to these other great divides. The opposition between externalism and internalism creates its own intellectual divisions and can be known by its own characteristic rhetoric and pattern.

Two large-scale externalist programs, which we will return to often throughout this book, are *adaptationism* in biology and *empiricism* in epistemology. In adaptationism the externalist pattern of explanation is displayed more clearly than it is anywhere else. Adaptationism seeks to explain the structure and behavior of biological systems in terms of pressures and requirements imposed by the systems' environments. Biological structure, or some very significant portion of it, is understood as an adaptive response to environmental conditions. The primary mechanism for this adaptive response recognized today is natural selection on genetic variants; genes are a channel through which the environment speaks. But this is not the only way to have an adaptationist biology. Other mechanisms, both natural and supernatural, can establish a relationship of adaptation or "fit" between organism and environment.

The case of empiricism is more complicated, as empiricism has come in so many varieties and has been used to so many different ends. The empiricist work I take as paradigmatic here is that of figures like Locke, Berkeley, Hume and J. S. Mill. What is central is a picture of how thought works and what it does, rather than (for example) an ideal of epistemic caution or a wariness of unobserved entities; we are mainly concerned with the psychological and descriptive side of the empiricist project. The central empiricist claim in this context is the claim that the contents of thought are determined, directed or strongly constrained by the properties of experience. In strong forms, this is the claim that there is nothing in the mind that was not previously in sense.

It will be apparent that when I say externalist explanations explain organic properties in terms of environmental properties, and then say empiricism is externalist, I understand "environment" and also "organic system" in a very inclusive way. Empiricists such as Locke, Hume and Mill were generally wary of theorizing about specific properties of the external world that might explain particular aspects of thought. They start the story at the interface between mind and world, and give a theory about what happens to agents under the impact of experience.[2] So experience might be the "environment" for the mind, if the empiricism of someone like Hume is to be understood as an externalist view.

A problem is posed by the fact that experience is usually understood, by empiricists and many others, in such a way that its characteristics depend on the properties of perceptual mechanisms. So this "environment for thought" is not fully external to thought. In Chapter 1 I said that one of the distinctive aspects of Spencer's and Dewey's views of the mind is their willingness to begin their account of cognition in specific properties of the external environment. It is important not to lose sight of this difference between Spencer and Dewey on the one hand and more familiar empiricisms on the other. But it is important also to recognize the ways in which more orthodox empiricist views do exemplify a basically externalist pattern of explanation.

It is characteristic of mainstream empiricism to think that experience has its own intrinsic patterns, to which the mind responds via mechanisms such as association. A "constant conjunction" of two properties in experience, for example, is not our imposition but rather something we *find*. We adjust to these patterns but do not create them, just as evolution produces an adjustment between the colors of moths and the patterns in their surrounds. The idea that experience has intrinsic patterns which it is the role of the mind to track or adjust to might be seen as an assumption which has been challenged, in different ways and with different agendas, by writers as diverse as Kant (1781), William James (1890), Nelson Goodman (1955) and Karl Popper (1959).

So, if we want to view empiricism in the style of Locke and Hume as externalist, we can do this in two ways. We could regard the "organic system" whose nature is to be explained as that part of the mind that lies inside the circle of perception, the agent who observes the "inner movie screen" on which sense-impressions are presented. Alternately we can regard empiricism as comprising just a *part* of a fully externalist account of the mind, with another part of the story (the origins of experience in external things) untold and perhaps untellable.

In addition to individualist forms of empiricism, allied programs in the philosophy of science exemplify an externalist pattern of explanation. (Here my terminology becomes very unorthodox.) Empiricist philosophy of science takes as its explanandum the system of scientific belief, and perhaps also the system of scientific practice.

Again it will be clear that I use the term "organic system" very broadly in my definition of externalism. Organic systems include individual organisms, minds, biological populations, and also social entities. My use of this term to denote social systems is not meant to imply what is sometimes called an "organicist" conception of social structure, or a biological approach to social science. Organic systems in the present sense need not themselves be alive, although they must perhaps be composed, at least in part, of things that are alive. The term "organic" does have a number of inappropriate associations – as does "system," for that matter. But "organic system" is used in a very loose and general way in this book, and no allusion to "general systems theory" or the like is intended. The term is meant to cover biological, intelligent and social entities without implying any view about the relations between these domains.

For the empiricist philosopher of science, observations (or observation reports) play a special role in the explanation of properties of the system of scientific belief and practice. In some empiricist pictures, observation may be conceived as an interface between the system of scientific belief and an autonomous external realm; there is great pressure exerted on the system of scientific belief by events occurring at this interface, and to predict the trajectory taken by the system of scientific belief we do best to attend to events and patterns found at this interface. Scientific theorizing responds to these patterns, or accommodates them, without creating them.

There have been empiricist philosophers of science who have explicitly resisted the idea that the data science deals with is given from without, that experience is an "interface" or channel of any kind. Moritz Schlick of the Vienna Circle is an example (1932/33, pp. 38–40), and logical positivists in general would say that such a view of the situation is metaphysical and meaningless. They would not deny, however, that data or "the given" does contain intrinsic patterns which theory is designed to accommodate; the task of science, for Schlick, is discovering and describing regularities in the given.

I am uncertain about exactly how far to extend the analysis of empiricism as an externalist doctrine, and this problem is especially difficult in the domain of philosophy of science.[3] The influential "constructive empiricism" of Van Fraassen (1980) is another empiricist view which avoids a strongly externalist picture. It is the *psychologistic* versions of empiricism which most clearly exemplify externalism; empiricism as a whole is externalist to the extent that the psychological picture is basic to empiricism.

It is important also that not all "orthodox" or "traditional" views of science are externalist. Within this family of views, it is the *data*-driven conception of scientific activity that has this status. If science is regarded as driven more by the internal development of theory, or the exploration of deductive relationships between postulates, this is not an externalist view. Further, the externalist about science need not be a scientific realist; he or she does not have to believe that through tracking patterns in experience we are able to attain knowledge of an external realm. It may be that, due to the limitations inherent in our epistemic position, all we can hope to do is describe patterns in experience without justifiably referring these patterns to something else.

Like empiricism, adaptationism has both individualistic versions and versions that take population-level properties as explananda. For example, some evolutionists have sought to explain genetic complexity in populations in terms of complexity in the populations' environments. Here the population is the organic system and the explanation is c-externalist. We will look at this case more closely in Chapter 9.

There have been several distinct externalist programs in psychology. The obvious examples are the various forms of *associationism*. "Association of ideas" was a mechanism for explaining complex thought closely allied with empiricism. Spencer, for example, explained the degree of cohesion or correlation between any two ideas in terms of the degree of correlation between the objects the ideas denote – Spencer's "law of intelligence," which will be discussed in Chapter 3. Associationism includes not just theories expressed in terms of ideas, but also behaviorist learning theory. The behaviorist rejects mentalist language but explains the structure of an organism's behavior in terms of patterns existing in stimuli which it has received, or patterns in the schedule of reinforcement on which it has been trained. Some research into "connectionist" models of thought, in which networks of neuron-like elements are trained to detect patterns in "training sets" of stimuli, is associationist – and externalist – in the same sense.[4]

J. J. Gibson's "direct" or "ecological" approach to perception (1966, 1979) can also be regarded as an externalist view in psychology. Gibson opposed the idea that perception must be explained in terms of an inference from meager data. He claimed that the information contained in light is sufficient to explain visual perception, once a realistic conception of the observer's activities and location in the world is assumed.[5]

Gibson's psychology and behaviorist learning theory are both opposed to psychological models based on processes of inference. They stress the richness and specificity of inputs coming into the system from outside, and are opposed to explanations that place great weight on details of the internal structure of the mind. This is not to say that there is a necessary link between an externalist perspective and a distrust of inner mechanisms. A cognitivist psychology can be externalist if it explains inner mechanisms in a strongly adaptationist way. This picture was embraced by Herbert Spencer, and this is also the form of some more recent externalist approaches in psychology (Shepard, 1987; Cosmides and Tooby, 1987).

The battles between behaviorism and cognitive psychology during the third quarter of the twentieth century were *partly* battles over the value of the externalist approach, but they cannot be understood solely in these terms. Some of Chomsky's arguments against behaviorist views of language are focused directly on the question of externalism, and the need for an understanding of the specificities of inner structure when explaining the output of a complex system (Chomsky, 1956). In particular, externalist views of the initial acquisition of language by children fail because there is too much organized complexity in the speaker's output to be explained in terms of the meager input a child receives. This is Chomsky's celebrated argument from the "poverty of the stimulus."

Though one aspect of the "cognitivist" movement was anti-externalist, another part of the revolt against behaviorism had to do with throwing off the burden of an overly restrictive philosophy of science, which had an excessive distrust of theoretical entities. The Spencerian externalist can make camp with the MIT internalist in this campaign. Further, although some central cognitivist work is focused on establishing an explanatory role for specific internal mechanisms, there has always been a more externalist camp within cognitivism as well. Herbert Simon's influential book *The Sciences of the Artificial* (1981) is overtly externalist. Simon defines his subject matter, "the artificial," as things which "are as they are only because of a system's being moulded, by goals or purposes, to the environment in which it lives" (1981, p. ix). This adaptation is often constrained by the capacities of the system itself; "satisficing," or producing an acceptable solution, must in many cases replace optimization as a goal. But in Simon's sciences of the artificial, which include psychology, internal factors only play this subsidiary constraining role. The basic processes studied are those whereby adaptive systems seek to "mould themselves to the shape of the task environment" (1981, p. 97).

Besides empiricism, there are more recent and local examples of externalist programs in philosophy. In contemporary philosophy of mind there is a range of naturalistic theories of meaning that make use of causal relations and reliable correlations between inner states and the world to explain the semantic content of thought. The most popular approach is the "indicator" or information-based approach of Dretske (1981, 1988), Fodor (1987) and others. Indicator semantics is a close relative of the epistemological position "reliabilism" (Goldman, 1986), which uses relations of this type to explain knowledge and justification, rather than content. These are both externalist views.[6] Indicator semantics and reliabilism are both opposed by internalist theories – indicator semantics is opposed by the "conceptual role" approach (Block, 1986), and reliabilism by "coherentist" epistemology (Lehrer, 1974). It is an externalist faith that is expressed by Dretske (1983), who said in response to the conceptual-role view of Paul Churchland, that no matter how hard the crank on the machine is turned, you will not get ice cream out unless you put in some cream. It is externalism that is expressed by evolutionists who admit that there are various kinds of causal explanations possible for a trait, but hold that the external, ecological factor that exerted selective pressure is the "ultimate cause" (Mayr, 1961).[7]

2.3 Internalism

Adaptationism and empiricism are examples of externalist views. An examination of opposition to these two programs furnishes good illustrations of internalism.

Internalist positions in biology, and internalist criticism of adaptationism, take many forms. Internalists claim that it is not possible to explain the structure of organic systems, or the course of evolution, by attending simply to the structure of the environments which organisms inhabit. Organic structure is strongly constrained, and in some cases determined, by internal factors, associated with the integrated nature of living systems. The properties of individual development from egg to adult, in particular, have long been thought to direct or constrain evolutionary change in this way. A strong internalism in biology is also seen in the search for ahistorical "laws of form" undertaken by the "rational morphology" tradition, associated with such figures as Geoffroy Saint-Hilaire and Richard Owen. This tradition stresses the underlying order and unity found in biological systems (Coleman, 1971). A contemporary view of this type is the "structuralism" defended by Brian Goodwin and others (Webster and Goodwin, 1982; Goodwin, 1986).

Several lines of thought within genetics are internalist, and have been used specifically to oppose more adaptationist, externalist conceptions of evolution. This has a long history. Much of the earliest work in genetics was associated with evolutionary views which rejected or downplayed the mechanism of natural selection. The man who coined the term "genetics," William Bateson, was strongly opposed to the Darwinism of his day. Bateson and also Hugo de Vries, one of the men who brought Mendel's work to the attention of science around the turn of the twentieth century, were among those who favored "saltationist" views of evolution. Saltationism holds that new biological forms can be produced in sudden jumps. On this view, mutation itself is the chief source of novelty and diversity in life. This was one of a range of internalist alternatives to Darwinism around the beginning of the twentieth century (see Provine, 1971; Bowler, 1989).

I also count as internalist certain lines of argument in contemporary population genetics, which investigate the relation between natural selection and the internal properties of genetic systems. Some geneticists have long criticized adaptationists by saying that the results of natural selection depend heavily on details of the genetic system and population structure (see Kitcher, 1985, Chapter 7, for an accessible review). If the best trait is coded by a heterozygote, which has two different genes at a given place on each of its two sets of chromosomes, then selection cannot "fix" that trait in the population. There will always be some individuals without it, at least at the juvenile stage.

The language of "constraint" in evolutionary biology is often the language of a moderate internalism, or of concession to internalist arguments.[8] A stronger internalism than this is found in work on internal processes that are held to determine or contribute to the trajectory of evolutionary change, rather than simply block or obstruct tracking of the environment. The "neutral theory" of molecular evolution (Kimura, 1983) is specifically opposed to the idea that observed patterns of genetic variation at the molecular level should be explained in terms of natural selection. Neutralism contends that random genetic drift accounts for the properties of the bulk of molecular genetic variation. Drift is a consequence of the finite size of natural populations, and its specific dynamic properties are a consequence of internal genetic details of the population in question.

One might accept the neutral theory but reply that if natural selection can "see" a gene, and the population size is large enough, selection will then determine its fate. Other workers in genetics have sought to show how the trajectory taken by an evolving system can be strongly conditioned by properties of linkage and recombination of genes, and the details of sexual reproduction (Lewontin, 1974a). In extreme cases these factors can lead to evolution taking a trajectory opposed to that predicted by selection – mean fitness in the population can decline (Kojima and Kelleher, 1961). The importance of such internal factors is the subject of ongoing debate in evolutionary theory.[9]

Another interesting internalist evolutionary mechanism is found in Jean-Baptiste Lamarck's *Zoological Philosophy* (1809).[10] Lamarck is remembered today mainly for his claim that evolution can occur as a consequence of the "inheritance of acquired characteristics." According to this view, structural modifications to organisms are the consequence of changes in habits, changes made by organisms in response to their environments. These structural modifications can be inherited, and they accumulate to produce large-scale change. That is how we usually think of Lamarck, and this mechanism is often described as "Lamarckian." But this was just one of Lamarck's evolutionary mechanisms. The other mechanism is an intrinsic property of life which constantly generates increases in the complexity of organization. Lamarck saw life as dependent on the actions of fluids that pervade all living things. These fluids course through the body, carving new channels through tissues as a consequence. The result is the creation of new organs, and a steady increase in the complexity of organization over time. These fluids include both familiar "containable" organic fluids, and, more importantly, "subtle" and "uncontainable" fluids such as heat (caloric) and electricity, which are fundamental to all living activity. Some of these fluids exist in the environment as well as in organisms, and in the simplest animals (such as jellyfish) the fluids act on the organism from outside. But once we reach moderately complex organisms the fluids are largely internally produced and maintained. They are what gives rise to the "inner feeling" which directs behavior in more complex organisms.[11] The containable and uncontainable fluids have their own dynamic properties of motion through the organism, and these properties bring about the steady increase in organic complexity. This part of Lamarck's view is primarily internalist.[12]

Internalist opposition to empiricism, of course, has a long history. Clear examples are provided by forms of "rationalism," such as the view of Leibniz, who criticized Locke's empiricism and proposed that we view the mind not as a "tabula rasa" (blank slate), but rather as pre-structured in something like the way that a piece of marble to be carved into a statue is pre-structured by its veins. The marble needs work before it takes on a statue's shape, but owing to the veins it will more readily take on one shape than another (Leibniz, 1765, p. 52). Internalists claim that there is no way ideas which come into the mind from outside can be formed into beliefs and judgements without the operation of specific internal mechanisms. Inputs will not just coalesce into beliefs.

Chomsky's "poverty of the stimulus" argument, which was mentioned in the previous section, is one paradigm pattern of attack on externalism: there is not enough structure in what is coming in to the system from outside to explain the structure in the system and its output. Further, though Chomsky holds that language acquisition is explained partly in terms of innate structure in the mind, he does not compromise this internalism by accepting an adaptationist explanation of this innate structure. Chomsky has been consistently skeptical about the project of giving an adaptationist explanation for the structure of language, stressing the possibility that the operation of physical laws on the organization of the brain, and factors associated with molecular biology, are more important than natural selection in this case (1980, p. 100; 1988, pp. 167, 170).

Here Chomsky raises externalist eyebrows. Externalists often concede rich inner structure, but then hold out for an adaptationist explanation of this structure, thereby achieving "externalism at one remove." From this viewpoint Chomsky's nativism looks like "buck-passing" (Dennett, 1980). This, as we will see in Chapter 3, was Spencer's reply to Kant's arguments about the need for prior internal structure in the mind.

Dennett (1995) uses a related line of argument against some internalist views in the philosophy of biology as well. The work of Stuart Kauffman (1993) on the interaction of development and selection in evolution is often regarded as supporting a (moderate) internalist view of some sort. Dennett, however, claims that what Kauffman and many others who work on the internal properties of evolutionary mechanisms are doing is "meta-engineering." The internal factors they describe are not "laws of form" or other inevitable properties of biological systems, but very early products of evolution, which have a selective rationale in the usual sense. Some are products of selection for evolvability and adaptability.

It might be thought misleading to use the term "externalist" for any view which concedes an important role to inner structure, even if the inner structure is explained in externalist terms. The explanation of the inner structure is, after all, a different explanation from the explanation of the behavior caused *by* that inner structure. This is right, but there is a variety of attitudes that can be taken to the two explanations. The views that I see as strongly externalist in this context are the ones that regard the inner structure in question as a mere waystation, as the means by which the environment exerts its influence. Dennett's claim that Chomsky is "buck-passing" falls into this category, as do views that regard the evolutionary rationale for an inner mechanism as in some sense the "ultimate" cause for behavior the mechanism produces.[13]

The poverty of the stimulus argument is one basic type of attack on externalism. Another attack, which I associated in the previous section with Kant and Goodman, denies that we can view the operations of the mind as tracking or adapting to patterns given in experience. It claims that these patterns are imposed or created, and depend on the operation of our categories and concepts. This line of attack on externalism is particularly relevant to the environmental complexity thesis, as it challenges the idea that environments can possess objective properties of complexity or variability to which organisms can be seen as adapting. This issue will be discussed in detail in Chapter 5.

I also understand much of Kuhn's *Structure of Scientific Revolutions* (1962) as an internalist critique of a certain empiricist picture of science. This empiricist picture views science as controlled by events occurring at an observational interface with the external world. Kuhn claimed that this account is descriptively inadequate as a consequence of its neglect of the role played by properties of the scientific community itself. Empiricism neglects the role played by scientists' theoretical expectations (which affect observation), standards of evidence (which are not derived from without) and science's intrinsic community structure (which affects which "inputs" will be attended to and which will be ignored).

Some workers influenced by Kuhn replaced the external constraint of experiment and observation with another type of external factor, the "interests" of the social groups and classes to which scientists belong. This is one form of sociology of science, sometimes called the "Edinburgh school." Not all the "interests" and social facts used by these sociologists are properly regarded as external. For example, differences in the scientific skills possessed by agents are often used to explain choices. But in many cases there is also an appeal to broader social and political, or "macro-sociological," factors. This is externalism.[14]

The opposition between those seeking to describe the roles social and political factors play in scientific change, and those defending more orthodox, empiricist positions, led to the proliferation of a set of terms which are ironic (to say the least) from my point of view, and which I think have muddied the waters. Experiment was taken to be a factor "internal" to science, and the political interests of a social class were viewed as "external" to science. On my view both empiricism and the macro-sociological forms of "interest theory" are externalist explanatory programs.[15]

In more recent years another genuinely internalist perspective has appeared in the field of science studies, in this respect the heir to Kuhn. To some extent this view has displaced accounts based on the "interests" of social groups. This is the type of sociology of science exemplified by Latour and Woolgar's *Laboratory Life* (1986). For Latour and Woolgar the internal economy of science has its own principles, its own currency, and the construction of a scientific fact is not controlled either by the outcomes of experiments or by factors deriving from political life at large.

Latour's subsequent work has continued to reject externalist views. In his *Science in Action* (1987) a rejection of externalist approaches is codified explicitly in the form of "Rules of Method." Rule 3 instructs us not to explain the settlement of a scientific controversy in terms of "Nature" pushing us one way or the other. Rule 4 instructs us not to explain it in terms of the influence of "Society" either (1987, p. 258). Latour's rejection of externalism of both types is clear. Latour's relation to internalism will be discussed in a later section of this chapter.

There may be more abstract oppositions that can be understood in terms of the divide between externalism and internalism. The very idea of taking language as a system of signs that refer to things outside the system may be said to be externalist (in a broader sense than the official sense of this book). In this case, and in some others, the fact that an idea is associated with the externalist camp is suggested by the distinctive ways in which it is opposed by internalists. Strong forms of structuralism in linguistics are the relevant internalist opposition here.[16] Saussure's famous *Course in General Linguistics* (1910), for example, is internalist both in its rejection of a referential approach to semantics (p. 65) and in its focus on the internal structure of language as opposed to its use in speech.

2.4 The larger landscape

I have been trying to sketch a picture in which the distinction between externalist and internalist approaches to understanding organic systems is a basic single feature of the intellectual landscape. But this is not the *only* basic feature of the landscape. Sometimes when large-scale battle lines are perceived in intellectual life, what I am calling externalism is grouped with reductionism, and is seen as part of one larger entity; an atomistic, or scientistic, "western" approach to inquiry. Externalism is also part of the "Cartesian reductionism" of Lewontin (1977, 1991). Similarly, according to Dahlbom (1985) most of the views I regard as externalist are part of a single "enlightenment" explanatory strategy, which is opposed by "romanticism." Others might seek to align the externalism/internalism divide with James' distinction between the "tough and the tender" in philosophy (1907). I think that although the externalism/ internalism divide can interact in various ways with other large-scale debates, it cannot be reduced to some other more fundamental opposition.

Firstly, the scientific projects of Chomsky, Saussure and Kimura are not at all anti-enlightenment (or especially tender-minded). But their views are just as internalist, within their respective fields, as more "romantic" approaches. All three assert the importance of internal causes and structures, and their claims about these internal causes constitute science in the enlightenment tradition.[17]

The new interdisciplinary field known simply as "complexity" is in large part founded on the conviction that an approach to explaining complex structure and behavior has been found that is rigorous and scientific, but which is not based on a picture of organic systems developing as functions of their environments (Lewin, 1992; Kauffman, 1993). Workers in this field attempt to find abstract mathematical properties of systems that tend to generate complex organization and behavior. Organization is largely explained as *self*-organization. Some recent work on "artificial life" realized in computer programs has the same character (Langton, 1989). "Cellular automata," for example, are mathematically characterized patterns in space that change over time in ways that preserve characteristic organizational features, without interacting with a structured environment. Research on cellular automata exemplifies the internalist side of this work most clearly; there is *no* concept of environment in this work, other than the space in which the system resides.[18]

The divide between reductionists and their opponents is also orthogonal to that between internalism and externalism. This is especially clear in the case of biology. In biology the relationship between reductionist and externalist (adaptationist) views has at many times been one of opposition. One of the basic *alternatives* to an externalist explanation of some specific aspect of an organic system is to inquire into the internal mechanisms that make the thing work the way it does. Rather than looking outside for an environmental rationale, one looks to the nuts and bolts inside the system. It has taken a large literature to establish that mechanistic and teleonomic explanations are not in competition in biology (Mayr, 1961; Tinbergen, 1963). Reductionism is a tough-minded but often internalist approach.

So internalism sometimes has a romantic, holist or idealist quality, and sometimes it does not. The debate between externalism and its opponents plays its own role in intellectual life, and choices made about externalism leave open other decisions about reduction, explanation and the role of reason.

There are reasons to be cautious about the perception of very large-scale structures in the intellectual landscape. On the other hand, some mention must be made of a fact about the relation between the intellectual landscape and one aspect of the *physical* landscape. A feature of the list of names and "isms" in this chapter that surely stands out is a strong tendency for externalists to be English. Adaptationism and empiricism, which I take as paradigm externalist views, are both thoroughly English.[19] Both elicit much shaking of heads in continental Europe, and mixed although more positive feelings in the U.S. In looking for internalists, Europe is often a good place to start, and several of the internalists I have discussed have been French.[20]

Of course, here again there is a risk of bad oversimplification, forgetting externalist Continental philosophers of the Enlightenment such as Condillac and Helvétius, and internalist biologists William Bateson and Brian Goodwin in England. But the overall correlation between nationality and allegiance might be taken as some support for the idea that there is a single externalist explanatory strategy, and a unitary internalist *bauplan*, which are encountered in similar forms in a variety of fields.[21]

Where there is similarity of form it can be the product of convergence or constraint, but it can also be the product of common descent. What are the historical roots of externalist patterns of explanation, and internalist opposition?

I am not a knowledgeable enough historian to make definite claims here, and I will not discuss connections to ancient or medieval thought. Certainly Locke is an important figure in the story. A range of Enlightenment figures influenced by Locke also developed strongly optimistic views about the perfectibility of man through the manipulation of his conditions, which spread widely through intellectual life in the eighteenth century (Helvétius, 1773).[22] Such a faith was expressed in the early nineteenth century by the utopian socialist Robert Owen in a well-known phrase: "Man is the creature of circumstances."[23]

Within biology there may be a separate lineage of externalist thought associated with the tradition of "Natural Theology." This tradition attempted to reveal God's work in His creation by describing the extreme adaptation of organisms to their stations in life. Many British biologists, including Darwin, were strongly influenced by this body of scientific and religious work. John Ray published an important work in this tradition in 1691, just around the time of Locke's *Essay* (1690). Natural Theology reached a peak in its influence with Paley's work (1802), and his classic "watchmaker" formulation of the Argument from Design.

Externalist evolutionary thinking came later. Many early evolutionists did give some role to the environment (see Bowler, 1989, for a survey). Lamarck, for example, had a mixed view which made use of separate internal and external factors. More thoroughgoing externalist evolutionary views appeared with Darwin, Wallace and Spencer.

Tracing the lineage of externalist thought in the history of science and philosophy is an important project, but not one for this book. Accordingly, in this work I understand the categories of externalism and internalism in what biologists would call a "phenetic" way. Explanatory programs are grouped by similarity rather than by descent. David Hull (1988) has urged a more "cladistic," descent-based approach to categorization in the history of ideas. I agree with Hull about the importance of tracing the historical lineage. But that is a further task, and I also think that a phenetic classification has its own value here. For example, there is value in inquiring into the logic and pattern of "reductionist" programs of explanation, even though it is often hard to work out who exactly is a reductionist, and whether or not reductionism forms a single historical lineage. My attitude to externalism is analogous.

2.5 Contesting the explanandum

Now the broad outlines of the opposition between internalism and externalism have been laid out, we will start to look at some more detailed aspects of the landscape, and at philosophical issues which arise when we try to understand particular debates in these terms.

In extreme cases, such as the views of Spencer and Chomsky, we find packages of commitments that reflect overall externalist or internalist outlooks on the world. But clearly one need not be an externalist or an internalist about everything. One might take one view in psychology and another in biology. Even within a specific field one might prefer different approaches to slightly different phenomena.

Biology contains interesting cases of "mixed" views. Stephen Gould, in a paper that surveys internalist and externalist views of the history of life, opts for a position of this type (Gould, 1977). Gould says that his generally environment-driven view of evolution is tempered by a belief that large-scale macro-evolutionary patterns have their own intrinsic dynamic, of the sort associated with "internalist" views. I am not sure if Gould would accept the label "environmentalist" for his present view of micro-evolutionary change. But the view Gould expressed in this 1977 paper illustrates the possibility of clearly defined mixtures.

Another mixture is found in Lamarck (1809). Lamarck, as outlined earlier, explained the overall increase in complexity observed in evolution with a largely internalist mechanism involving inner fluids. But he accepted that this increase in complexity is an imperfect pattern in the biological world. The effects of the environment are used to explain the residue. Externalist arguments are used to explain departures from an orderly progression, with respect to complexity, in the development of life. When organisms find themselves in particular environments and adapt to them, they are led off the path directed by the internal engine of evolution. So Lamarck distinguishes between two types of explanandum in evolution – the overall increase in complexity, and the deviations from it – and makes use of internalist and externalist mechanisms to deal with each, respectively.

These cases suggest that some internalists and externalists might be able to make peace by recognizing that subtle differences between explananda could lie behind their differences in preferred mechanisms. Perhaps this is so. But it is

also important to realize that many debates between internalists and externalists are carried on as debates over what should be recognized as important
explananda. This fact is of immediate relevance to us because *complexity* is
a favorite explanandum of externalists. It is characteristic of some externalism
to argue that complexity poses such deep or important problems for science
that the mechanisms which explain complexity acquire some special pre-
eminence.

This is clear in the case of adaptationism, and well illustrated by Dawkins'
The Blind Watchmaker (1986). In the first pages of Dawkins' book organized
complexity is proclaimed to be *the* problem which motivates biological
inquiry, and marks the subject matter of biology off from physics. Evolution
by natural selection – an externalist mechanism – is *the* answer. Other mechanisms in evolution, such as random drift, may explain what appears to be
a large bulk of what is observed in the world. But they are in principle unable
to solve the most important problems.

I think this expresses a common response by adaptationists to the criticisms
of people like Gould and Lewontin (1978). The reply is not that natural
selection is the only factor in evolution (a view no one holds), but that selection
is the only evolutionary force with a certain type of explanatory capacity. It is
the evolutionary factor that can explain complex adaptation. Through the
privileging of a certain explanandum, a special place is retained for the
externalist explanation.

It may be that the distinctive role played by assumptions of parsimony
(simplicity) in some empiricist epistemology and philosophy of science can be
understood in the same way. If parsimony is accepted as a constraint on good
reasoning or good science, then the default assumption is a small ontology or
a simple view of the world. When a system of belief is not perturbed it does not
grow with respect to its overall or "net" set of commitments about the world.
For the empiricist, a good scientist has to be induced by experimental results to
add new complications to their theory of what is going on. Adding a new
particle to a theory is supposed to be the result of intrinsic complexities in
experimental results that cannot be accommodated in any other way. So
a modest set of commitments does not need much explanation. What has to be
explained is the dazzling complexity in a paleontologist's or astro-physicist's
world view. Why are they committed to the existence of *all those things*?

In other ways as well we can see empiricists replying to critics by privileging
a particular explanandum. The beleaguered empiricist can make moves
analogous to that made by Dawkins in defence of adaptationism. An empiricist might accept that science in practice has many of the properties that Kuhn

and others ascribe to it, and accept that many aspects of the course taken by scientific belief are determined by factors other than observational data. But the empiricist can then insist that some special feature of science, perhaps its success (on some measure) or the convergence scientific belief displays over time, is to be explained in terms of the impact of data.[24]

There are several avenues of reply available here to internalists (or at least, to anti-externalists). One is to object to this privileging of a particular explanandum. What is so special about complexity? Why should we not just seek to explain most of what we see, and treat it all on a par? Or at least, why not treat it all on a par until some special reason is given to elevate some phenomena and disregard others? So when someone like Dawkins says that although drift might explain many observations of molecular genetic variation, it is useless in addressing the big questions about adaptive complexity, an advocate of genetic drift such as Kimura can object: why are the molecular genetic phenomena not worthy of attention and explanation? "[W]hat is important in science is to find out the truth, so the neutral theory should be of value if it is valid as a scientific hypothesis" (Kimura, 1983, p. 325).[25] The privileging of a certain explanandum can be contested whether or not the reality of the phenomenon is denied.

Internalists can also mount something stronger than this "democratic" objection. They can claim that some *other* phenomena are actually the important ones. I said that I view Latour and Woolgar's *Laboratory Life* as an internalist work in the sociology of science, directed both against empiricism and also other externalist positions. Latour and Woolgar do focus on a particular explanandum in their discussion. But this is the *order* displayed by the system of scientific belief and practice (1986, Chapter 6). They want to explain unity and stability, not complexity. I think it is fair to say that many internalists in biology, especially those concerned with development and similar fields, think that the order and the unity of organic systems are phenomena underestimated by externalists, and phenomena that externalists do not have good resources for dealing with (see Kauffman, 1993, for example). For some internalists, a large proportion of the externalist's much-vaunted "complexity" might be just noise, or at most, bells and whistles. It is not something that demands pride of place as a phenomenon in science.

Having said this, I must concede that Chomsky, one of the best examples of a thoroughgoing internalist, flies in the face of this pattern. Chomsky's arguments against behaviorism focus precisely on complexity as a special and vexing explanandum. Complexity in a mature speaker's language use is a real phenomenon, and one that demands a specific type of explanation.[26] Second, the interdisciplinary field known as "complexity" is very often internalist, as outlined in the previous section. It would be unwise to make this claim about favored explananda of externalists and internalists too strongly.

2.6 The location of the internal/external divide

I have been directly contrasting internalist and externalist perspectives on organic systems, as if there is always a clear prior division between the system and its surrounds, between organism and environment. When working out whether an explanatory program is externalist it is best to allow the field itself to determine what counts as internal and what counts as external. Sometimes there is consensus about this, or an obvious boundary, but of course this is also an issue that is often contested. In fact, a tactic used by some strongly internalist and externalist writers is to re-draw the boundary between system and environment in a way that makes the opposing view hard to sustain.

Herbert Simon's *The Sciences of the Artificial* is, I said earlier, a strongly externalist work although it advocates a cognitivist, information-processing approach to the mind. Thought is seen as an adaptive response to conditions in the environment. But surely a person's response to an environment encountered now is heavily dependent on prior experiences as embedded in memory? Surely we must place great stress on the internal factor of memory when explaining any intelligent agent's behavior? There are several ways for an externalist to respond. One option is to stress that whatever is in memory is the result of past interaction with the environment. Another way is to shift inward the boundary that determines where the environment starts. This is what Simon does: "I would like to view this information-packed memory less as part of the organism than as part of the environment to which it adapts" (1981, p. 65).

Daniel Dennett makes a similar move in a paper that defends the "law of effect" as a global and in some respects *a priori* principle for explaining behavior (1975). The law of effect claims, roughly, that actions which produce beneficial results tend to be repeated. Trial and error, for Dennett, is *the* mechanism behind learning and adaptation. Dennett agrees, however, that the mechanism of learning by reinforcement is not rich enough to explain the behavior of intelligent agents if it is required that the reinforcers used in the explanation are all "external" reinforcers such as food and electric shocks. Intelligent agents can plan and rehearse possibilities, shaping their behavior without direct interaction with the environment. But Dennett's response is to view this planning in terms of an internal process of trial and error, done with the aid of an "inner environment." Dennett is committed to the explanation of intelligence in terms of organism/environment interaction, so when something internal appears to play the role usually attributed to environment, this inner structure is labelled an "environment" against which variants are tested.[27]

This is not to say that nonstandard divisions between organic system and environment are used exclusively by externalists. Far from it. The view of life proposed by Humberto Maturana and Francisco Varela is an internalist one, in my sense (Maturana and Varela, 1973). They investigate sets of dynamical rules governing interactions between elements of systems, which result in the system being "self-producing." At one point they raise the issue of how this view treats the role of environmental events that seem important to any organic system's fate – environmental events the agent must respond to and adapt to, the bread and butter of externalist conceptions of cognition. Their response is to shift the boundaries of the organic system outward (1973, p. 78). If environmental features matter, then these too are internal to the system, not explanatorily important elements outside of it: all feedback is internal.

There are problems raised not just by nonstandard divisions into organism and environment, but also by cases in which it is hard to see what a "natural" division would be. I am taking adaptationism to be an example of an externalist program. But one of the most important bodies of work within broadly adaptationist approaches to biology in recent years has been evolutionary game theory (Maynard Smith, 1982). Evolutionary game theory seeks to explain certain patterns of animal behavior in terms of contests between individuals over resources. There is often a real environmental resource in the story – territories, food – but the details of the explanation depend specifically on the cost/benefit properties of different types of interactions between individuals, and these cost/benefit properties depend on what others in the population are doing. For example, it often does not pay to be aggressive if everyone else is aggressive, if everyone will fight hard in every contest. But if most of the population will only bluff and then retreat when challenged, aggression will be more successful. A mixture of behaviors is then predicted. Is this type of explanation externalist or not? The behavior of any individual is explained largely in terms of factors external to that individual. But if the population as a whole is the organic system in question, then its properties (the balance between aggressive and pacific behaviors, for example) are explained mostly in terms of factors internal to it.

In the context of the positive project of this book, game-theoretic explanations are externalist. Some types of cognition may be understood as devices that enable agents to deal with complex social environments, environments characterized by a range of behaviors that are best met with different responses. Game theory can be used to try to characterize social environments in which complex behavioral strategies are favored.[28] In other explanatory contexts, however, game-theoretic explanations need not be externalist.

More problems are created when a program of research takes as its aim the explanation of *the division between internal and external*. Earlier I gave Latour and Woolgar's *Laboratory Life* as an example of internalist research into science, and also noted the rejection of several forms of externalism in Latour's *Science in Action* (1987). In that work "Rule 3" forbids explanation in terms of the impress of data or facts; "Rule 4" forbids explanation in terms of the pressure of Society, and broader political-economic factors. But then there is "Rule 5," which instructs that we never take the inside/outside division for granted. Instead we should study the process whereby actors construct boundaries between science and the rest of society. We should follow the activities on both sides of any purported boundary of this sort (1987, p. 258). "Science" and "Society" are the *results* of this process whereby a boundary is set up and maintained.[29]

This is not straightforward internalism and I am not sure exactly what to do with it. The pattern Latour is describing is reminiscent of that described by some biologists, such as Maturana and Varela. A network of interacting elements "precipitate out" a distinct organic system with boundaries. The boundaries between organism and environment are actively maintained as a consequence of dynamic properties of the lower-level elements; the organic system is in this sense self-producing. For Latour a network of agents "precipitates out" science as a distinct entity by its construction of networks of support. The science/society boundary is like the membrane of a cell. I will leave this as a problem case, which might be linked more convincingly to my framework with further work.

2.7 Problems of adjudication

These different patterns of explanation have sometimes been described above as if they appear in completely pure forms, as if people want to place the *entire* explanatory burden on external factors, or internal. This is greatly oversimplifying, and when I call a view externalist, or internalist, there are two provisos which must be kept in mind.

First, the assignment of the label "externalist" is dependent to some extent on the other ideas which prevail in the relevant field. Against the background of behaviorist learning theory, Chomsky's views are strikingly internalist. But, of course, there is much that Chomsky explains in terms of external factors. Given that the English tend to speak English and the French speak French, there is only so much about language that can be explained internally. Lamarck and the population geneticists discussed earlier are similarly "internalist" only against a certain background. It is harder for me to think of someone who invites the label "externalist" only because of the strongly internalist views prevailing in their field (I am more used to externalism-dominated fields). This "relativity" of the categories I use is only partial; there is no reasonable set of background patterns of explanation against which Locke's view of the mind can fairly be called internalist.

The second proviso is related. When I say externalists place great weight on external factors and neglect internal factors, this does not imply that they deny that internal factors exist and are necessary in producing the behavior of the system they are interested in. A "genetics-free" adaptationist model in biology should not be viewed as denying that genes exist and that evolution depends on them. Any model abstracts, and focuses on some particular set of properties at the expense of others. If the theorist tried to build into their model every parameter they believed plays some role, the model would almost always be uselessly complex. An externalist, in my view, is not necessarily someone who believes that internal factors play no role. It is often someone who thinks that internal factors are less important than external; they think that attending to the details of what is going in to an organic system from outside will tell you more about the details of what it does than any other set of factors will.

Saying this does require a substantial philosophical assumption. I view an externalist in some field as a person who thinks external factors are "more important" or "more informative" than internal. But everyone agrees that, in almost all real systems, there will be *some* role played by both internal and external. Both internal and external factors are individually necessary, and neither is individually sufficient. The outcome is a consequence of the interaction of both factors. How are we then to decide when one type of factor plays "more" of a role than the other?

There are some cases in which specialized statistical measures can be used to work out the "proportion of variance explained" by each factor. But these measures are themselves controversial, if they are interpreted as giving causal information (Lewontin, 1974b), and in any case they can only be used in specific types of cases. They are not going to help us decide whether perception should be viewed as an inference, or whether population structure can be ignored in evolutionary models of behavior.

This problem raises an important possibility. Perhaps the idea of adjudicating between internalist and externalist perspectives on the mind, or on evolution, is nonsensical. Perhaps we are futilely trying to decompose phenomena which we know in our hearts cannot be decomposed in this way. We dislike attributing everything blankly to "an interaction between internal and external" and wish to give one side or other the larger prize or the greater blame. It is open to the reader to suspect that while I may have succeeded in the preceding sections in describing a certain rhetorical divide in the sciences and philosophy, with certain national or cultural traditions behind it, the debate itself is in most cases irrational or merely one over preferred rhetoric.

This view is related to the claims of psychologists and biologists who have argued that the idea of isolating internal *or* external factors as the causes that produce biological structure is misguided. Organic form, on this view, is always the product of an interaction between external stimulus and prior internal structure. If this interaction was taken seriously we would give up trying to give priority to either internal or external. We would recognize the interaction as having its *own* properties. Forms of this position can be seen in Piaget (1971), Waddington (1975) and Oyama (1985).[30]

Oyama's view is a challenging one here. She argues that the dispute between nativist/internalist and empiricist/externalist views of biological structure is the empty byproduct of a more fundamental mistake. This mistake is the idea that biological form must arise from the imprinting of some pre-existing information on inert matter – matter which is unable to organize itself without this help.

> When the problem of explaining the emergence and persistence of form arises on the biological stage, it is of minor consequence in the end whether the *deus ex machina* is called upon to create the living machine from the inside or from the outside (or whether there are two of them, each contributing certain components); the yoked ideas of pre-existing form and teleological control travel together. They require each other, imply each other, even conjure each other out of the wings if separated, exchange places in successive performances. (1985, p. 141)

Oyama thinks that, since most of us officially hold to a "canon of enlightened interactionism" about most organic systems, and accept that the extremes of total internal control and total external control are nonsense, the only reason that we continue to talk in terms of internalist and externalist "alternatives" is the fact that we cannot think about the genesis of biological form except in terms of information or a blueprint directing the process. If we give up this view of where biological form comes from, we can give up using the language which opposes internal and external as rivals.

Oyama's discussion is primarily concerned with explanations of individual development, and the opposition between nature and nurture in this area. I am not sure how generally she would make these claims.[31] But however broadly Oyama intends her own attack on the internalism/externalism dichotomy, the fact that I am accepting the dichotomy as real and useful requires that some reply be made to a general skepticism about apportioning responsibility between internal and external.

First, there is sometimes, in specific cases, a question of the existence of machinery. Either children start out life with very complicated domain-specific innate psychological mechanisms or they don't. Although some debates about internalism are debates about the existence and nature of machinery, in many cases there is no question as sharp as this that can decide the matter. The dispute is not about what exists, but about whether something makes a difference, and what kind of difference.

The position taken in this book is that being an externalist very often involves making an explanatory bet. Adaptationists bet that spending most of your time and energy investigating patterns of natural selection at the expense of genetic factors is better than spending more time and energy on the genetic system and less on patterns of selection. Adaptationism involves preferring having a model with four environmental parameters and two genetic parameters, to having a model with the investment switched. Empiricism is preferring knowledge of exactly which experiments a scientific community did, to knowledge of who was in charge of the journals and how they were refereed. Behaviorism is betting that knowing the patterns of reinforcement experienced by a set of learners will tell you more about what they do than knowing exactly how they are wired up will tell you. These attitudes all count as externalist, in the sense of this book.

So when confronted with a particular type of phenomenon – acquisition of grammar, the loss of toes in horses over evolutionary time, or whatever – researchers will tend to make a variety of explanatory bets. Some will design models in which external factors get most of the attention, and others will not. The amount of emphasis placed on the two types of factors can vary continuously. The same person could be inclined to make different bets with different subject matter.

If the competing models that deal with a particular phenomenon are comparably complex and well-developed, then we would hope to be able to ask: "which one fits the data better?" If an externalist model fits the data better, that is a (fallible) reason to believe that the phenomenon is due primarily to external causes; that external differences make more of a difference than internal ones do. When a great many particular cases within a general domain (such as evolution) fit an externalist pattern, that constitutes the success of externalism in that domain (see also Orzack and Sober, 1994).[32] We then have some reason to believe that external forces are more important than internal in that domain.

If there can be no agreement between internalists and externalists about how to evaluate the goodness of fit between the rival models and the data, then we have a case of incommensurability of standards, the most troublesome type of "incommensurability" in the philosophy of science (Kuhn, 1962). I do not have a general argument for the view that this incommensurability will not raise insoluble problems. In some cases it might. There are in fact two possible problems: (i) agreeing on the explanandum, and (ii) agreeing on the criteria for being an adequate model of the explanandum. Both have the potential to cause difficulties. Recall the earlier discussion of how the relative importance of different explananda is often contested by externalists and internalists. When internalist models achieve a very good fit in the modeling of some evolutionary phenomenon that matters to internalists, externalists may say that the explanandum is so trivial that no degree of success with it vindicates an internalist stance: "Let us see how you deal with a *real* problem" (as Dawkins said, in effect, to the neutralists). These are the hardest issues to resolve in an unbiased way.

On the other hand, there are cases in which both sides agree on what has to be explained. Both adaptationists and more "pluralist" evolutionary biologists would like good models of the circumstances in which populations exhibit unusual sex ratios (far more females than males, for example). The aim is to get a realistic model which predicts the various observed sex ratios in animals like wasps (Werren, 1980; Orzack, 1990). Then the question is whether both sides can agree on the criteria for having a model that fits the sex ratio data.

In this case it should be reasonable to expect that both sides could agree about what it is to get a reasonable fit between a predicted number (proportion of females) and the number observed. Not *everyone* in biology may agree about what a good fit between data and model is; the hope is rather that a commitment to one side or the other of this debate does not come with a commitment to what it is to have a model which fits the data on sex ratios. If the externalism/internalism divide *does* structure people's disputes about what a good fit to data is, then we do, admittedly, have a problem.

In any case, I will proceed on the assumption that externalists and internalists in various fields make explanatory bets, and that in many cases, at least, there is a real empirical question at issue. It is empirical whether or not sex ratio can be predicted with reasonable accuracy by an adaptationist model based on optimality reasoning.

One possibility that should be considered seriously is that there is usually a fact of the matter in any particular case where internalist and externalist models can be compared, but no fact of that matter about whether externalism or internalism is better in an entire general domain, such as "evolution." For here the incommensurability problems probably will be acute. Even if internalists and externalists manage to agree about who has dealt better with horse toes, with wasp sex ratios, with the extinction of dinosaurs, and so on, it may be very optimistic to hope to devise a test for who has explained "most" of evolution or "most of the important" evolutionary phenomena. Here I expect a certain degree of relativism must be accepted. Similarly, there is little that I can reply to someone who says: "Even if you can show that the environmental complexity thesis about cognition makes sense and explains what you say it does, it does not say anything about what *I* think is important about the mind."

The question of whether externalism (or internalism) is a coherent framework is not the same as the question of whether an externalist and an internalist can agree on a test that would adjudicate between their positions in a fair and neutral way. It may be that a commitment to an externalist outlook does at least come with a certain assignment of relative weights for different explananda. Some phenomena will matter more for an externalist's projects in understanding the world than they will to an internalist. The possibility of a neutral global test is desirable, but if it is unavailable that does not damn externalism as a coherent outlook. What would damn externalism is the possibility that the externalist's own scientific or philosophical beliefs imply that the idea of an externalist explanation is nonsensical given the degree of symmetry conceded between the roles played by internal and external. I interpret Oyama as thinking that many who use the language of external (or internal) control are in just this situation. It is this claim that my appeals to "explanatory bets," and choices made about what to include in a model when resources are limited, are intended to reply to.

I am viewing externalism as very often involving an explanatory bet. We should recognize however that some people who work in externalist programs will not have even this level of conviction. Choices made about what to include and what to leave out of a model may just reflect a desire to see how things turn out if a certain approach is taken (perhaps given that no one else is taking it), or a desire to see how much can be done with a specific method or strategy, or a desire to make use of a technique that the researcher has special expertise with. A theoretical program, considered in abstraction, can reflect a set of externalist bets, but people working in the program might just enjoy doing that particular sort of mathematics or field work. They might think it is more ethical to investigate what whole animals do in response to their environments than it is to take them apart. These people are not (thereby) externalists, though the programs they are working in are programs that reflect externalist bets.

At this point the reader may wonder what my own attitude is to externalism, and to the ongoing debates between externalism and internalism. Is this book a defence of the robustness of the externalist explanatory strategy, or an attempt to damn empiricism with its kinship to adaptationism (or vice versa)? In fact it is not written in praise or blame of externalism itself. The value of the externalist strategy can reasonably be expected to vary a great deal across different fields. With respect to the total cross-disciplinary program of externalism, my aim is to understand the program better. But this book is also an attempt to follow one particular externalist idea – the environmental complexity thesis about the function of mind – for some distance, because of the possibility that this idea can be the foundation for a good theory of the place of mind in nature. As stressed in the previous chapter, the environmental complexity thesis is a claim about the role played by certain core cognitive capacities; it does not require a strongly externalist view of the day-to-day operation of thought. So it does not, for example, require the truth of a strong form of empiricism.

The only types of externalism, and of anti-externalism, that are rejected here are the *a priori* kinds. These are claims that there is some completely general reason why externalist explanations can never work, or must always work, or must, objectively, be the best possible types of explanations. I include in this rejected category various *a priori* arguments that neither externalist nor internalist strategies can ever work, as everything must be attributed to the properties of both parties.

Each field I have discussed – and many others I have not discussed, especially in the social sciences – carries on its own opposition between externalist approaches and characteristic forms of dissent. The outcomes vary across fields. There is ongoing dispute about which phenomena to take as basic, which show the way forward to a deeper understanding of nature, which bets will pay off. Externalists caricature internalists as mystics and vitalists;

externalists in turn are caricatured as holding that the mind is a passive "mirror of nature," a *tabula rasa*, or as holding that every organism is a collection of atomic adaptive traits with no interactions between them. This trading of caricatures is as much a part of the process as the bets that the caricatures are intended to motivate.

Many fields feature an ongoing see-sawing between externalism and its rivals. When does a concession made by an externalist to internal factors constitute selling out or giving up? This depends on the field and is a matter of degree. Whenever externalist philosophical theories of meaning allow a larger place to the *how* the mind represents, concessions are made to internalists. Whenever evolutionists allow a larger place to the role of developmental programs in explaining the characteristics of organisms, these concessions are made again. Whenever a Kuhnian historian of science admits that experimental results "constrain" another aspect of the development of scientific belief and practice, concessions are made in the other direction.

My attempt to cast various debates as disputes between internalists and externalists should not be taken to deny that two truly externalist approaches can disagree, even when they are in basic agreement about the boundaries of the system and the phenomena to be explained. From the present perspective a striking thing about some of the "nature versus nurture" debate about human behavior as it developed through the twentieth century has been its partially in-house character. The relevant "nurture" view was often behaviorist, and the "nature" view was often (though not always) adaptationist. Spencer, as we will see, would have viewed much of it as a storm in his Victorian teacup. He would *not*, however, have seen Chomsky's assault on behaviorism as a storm in a teacup.

2.8 C-externalist explanations

The environmental complexity thesis is, on some interpretations, an externalist claim. In this book we will also examine a more specific category of externalist positions that some versions of the environmental complexity thesis exemplify. These are explanations of internal complexity in terms of environmental complexity: c-externalist explanations.

As complexity is a favored explanandum of externalists, many paradigm externalist explanations are c-externalist. Not all c-externalist views in different fields agree about what complexity is, however. As outlined in Chapter 1, in this book properties of complexity will be understood as properties of heterogeneity. This is the concept of complexity used in my positive discussions of the environmental complexity thesis. It is also reasonable to regard this as at least a component in the concept of complexity used in many other discussions.

On c-externalist views, environmental complexity explains internal complexity. What exactly is this "explanatory" relation? Different theories and subject matters will be associated with different conceptions of explanatory relationships. But one distinction is universally important here.

When I have presented some of this material in talks, I have twice been vigorously interrupted just after describing the idea of a c-externalist explanation, and saying I would talk about the idea of a general program of explaining organic complexity in environmental terms. Once at a talk to a cognitive science group, the person interrupting thought the idea behind these explanations was so obviously false that he could not see why I was putting it forward for discussion. After all, his cat received just the same environmental inputs that he did, if these inputs are described in any objective way. But the cat showed no signs of becoming as complex as him as a result. The other time I was interrupted, at a talk to people who do computer simulations of evolution, it was by someone who found the idea so obviously true that he could not understand why I was going to talk about it. What *else* would make things get more complex?

Part of the disagreement here (although only part of it) stems from interpreting the dependence involved in the program of c-externalist explanation in two different ways. Is the point that environmental complexity is *necessary* for organic complexity, or that environmental complexity is *sufficient* for organic complexity? There is a possible position we could call the "naive matching view": environmental complexity is both necessary and sufficient. Spencer, a c-externalist extremist, held something like this. His basic principle for explaining the complexity of thought, his "law of intelligence," was presented in that form (section 3.5 below). But this certainly is not the only view possible. It is more characteristic of reasonable c-externalist programs to say that environmental complexity is necessary but not sufficient for organic complexity, as I understand the literatures. Not all environmental complexity will lead to organic complexity – much environmental complexity can simply be ignored – but you will not get organic complexity in any other way.[33]

So when the cognitive scientist mentioned above objected that his cat made trouble for c-externalism, this can most readily be understood as a counter-example to the idea that environmental complexity is sufficient for organic complexity. The world is complex but the cat does not respond. Conversely, the person who saw the c-externalist program as obviously right may have conceived the claim primarily in the other direction. Though not all environmental complexity will lead to organic complexity, organic complexity is not produced in any other way, so you will not get organic complexity in a completely simple world.

The version of the c-externalist idea that asserts a necessary rather than sufficient link is not just more common, it also makes more sense. If an organic system was bound to attend to every type of complexity in the environment, then the complexity of most environments would be massive, indefinitely big. If survival required adjusting behavior to every complex pattern in the world, then in the short run we'd all be dead.

In this book we will discuss both claims that environmental complexity is necessary for organic complexity, and claims that it is sufficient. But claims about necessary connections will be more important.

2.9 Cognition as organic complexity

The environmental complexity thesis is on some interpretations a c-externalist approach to cognition.

It is certain *teleonomic* interpretations of the environmental complexity thesis that have this status. Cognition, as outlined in Chapter 1, will be viewed in this book as a means to behavioral complexity. When behavioral complexity is itself explained in terms of environmental complexity, that is a c-externalist explanation. If this link is in turn used to explain the complexity of cognitive capacities themselves, that is a c-externalist explanation also.

Not every teleonomic version of the environmental complexity thesis is necessarily c-externalist. Cognition is a complex organic capacity, but that is not all it is. Cognition also displays apparent properties of unity and order. If aspects of cognition other than its complexity are explained in terms of environmental complexity, then the explanation is externalist but not c-externalist.

When the environmental complexity thesis is interpreted in a purely instrumental way, saying nothing about why cognition exists, its status is more complicated. Then the explanation that cognition figures in is an explanation of the capacities of larger systems of which cognition is a component. For example, the capacity of the larger system in question might be the well-being of the whole organism, its getting by and staying alive. In that context, the environmental complexity thesis claims that cognition helps the organism to get by, through enabling it to deal effectively with environmental complexity. The explanandum is the organism getting by, which is not itself a property of organic complexity.

Another property of a "larger system" which cognition may contribute to is, ironically, *complexity*. Cognition, through its guidance of complex behavior, can contribute to the increasing complexity of certain organisms' environments. This is the converse of a c-externalist explanation; it is an explanation of environmental complexity in terms of organic complexity, a "c-constructive" explanation. This claim about cognition is not inconsistent with the environmental complexity thesis. It could well be that cognition helps us to cope with some types of environmental complexity, but does so by making other aspects of the environment more complex than they were before. Constructive arguments, and the transformation of the world via thought, will be discussed in Chapter 5.

The instrumental version of the environmental complexity thesis might still be regarded as "externalist" in a looser sense than the sense used here. The instrumental version of the thesis does stress the "outwards-directed" properties of mind, and asserts a relation between mind and external complexity. But the instrumental version of the environmental complexity thesis is not externalist in the sense used in this book, because the instrumental version of the thesis does not seek to *explain* the internal in terms of the external.

To summarize: insofar as the environmental complexity thesis is construed as an account of why organic systems have cognitive capacities, this thesis constitutes an externalist view. This is not to say that the environmental complexity thesis must be associated with an extreme form of externalism. Spencer's externalism was extreme, and Dewey's was very moderate, as we will see. Second, when the environmental complexity thesis is viewed as a claim about what cognition can do, what it is good for and what contribution it makes to other goings-on, it does not involve an externalist commitment.

Notes

1 My account of the opposition between internalism and externalism draws on several sources, and versions of this taxonomy have been presented before by writers with a variety of axes to grind. Spencer is, in a way, the man behind my version of the distinction because he embodies the externalist approach so comprehensively, in his fusion of adaptationism and empiricism. This is discussed in Chapter 3. Piaget (1971) discusses the analogy between strongly environment-driven views in biology and empiricist views of the mind, in the course of criticizing this general outlook. Aside from the basic point about the analogy, I disagree with many of the details of Piaget's account. Gould (1977) has a good account of externalist and internalist accounts of evolutionary change, with a focus on paleontology. For biology, see also Bowler (1989) and McShea (1991).

Dahlbom's version of the distinction (1985) is criticized in section 2.4, and see also Dennett (1987). Dahlbom opposes his categorization to that given in Dennett (1983). In that work Dennett asserted an analogy between behaviorism and anti-adaptationism of the Gould-and-Lewontin type. I agree with Dahlbom that the internalism/externalism divide (which sharply splits people like Gould from behaviorists) is more fundamental than the one Dennett describes. For psychology see also Fodor (1980).

Bechtel and Richardson (1993, Chapter 2) discuss the opposition between externalist and internalist views of the "control" of complex systems, including the oppositions between adaptationist and development-based views of evolution, and between behaviorism and Chomsky's mentalism. Oyama (1992) discusses internalist and externalist views of change in ontogeny and phylogeny, and makes some connections to the social sciences. She regards the internal/external distinction as an obstruction to our understanding of organic change. See also Latour's analysis of explanatory patterns in science studies (1992), although from my standpoint Latour's analysis does not properly distinguish internalist and constructive explanations.

2 Locke was prepared to make some general claims about the external world – that external objects have only primary qualities, for example. But he claimed that there could, in principle, be no explanation of how specific qualities in the external world produce particular simple ideas in us (1690, Book IV, Ch. 3, sec. 28–29). We have reason to believe that there are regular correlations between external qualities and the production of simple ideas, (Ch. 3, sec. 4), but if we inquire into the causal basis of these correlations, all we can do is attribute these facts to the will of the Wise Architect.

For Hume, the cause lying behind sensations is a matter "perfectly inexplicable by human reason." We can never know with certainty whether sensations are caused by external objects, by "the creative power of the mind," or by God (1739, Book I, Part 3, sec. 5). To the extent that the second option is a real possibility for Hume, his view is not properly externalist in my sense.

Berkeley, on the other hand, made specific claims about the role played by God in directing the course of ideas. See section 5.2 below.

3 There are complicating factors even in the paradigm cases. Locke accepted reflection on the workings of the mind as a genuine source of knowledge, and this is not externalist (as Houston Smit stressed to me). Hume included feelings of love and anger as "impressions," along with perceptions of apparently external things. And an important role was played in Hume's view by an internalist account of our concept of God.

Adaptationists can also explain one feature of an organism in terms of adaptation to another. My claims here concern the central, but not exclusive, focus of empiricist and adaptationist programs of explanation.

4 Rumelhart and McClelland (1986) do say that, strictly speaking, their "PDP" or connectionist approach is consistent with "rabid nativism," in which all the

connections between units which constitute an agent's knowledge are fixed at birth (1986, p. 139). It is also compatible, in this sense, with the most rabid empiricism. In general, work in the connectionist research program has stressed the power of simple, general-purpose learning rules that give rise to complex mental capacities when they interact with complex patterns in experience.

5 Part of what Gibson stressed was the active nature of the observer – moving around, looking from different angles and so on. But I view his position as externalist on the whole.

6 There are other overtly externalist views of meaning as well (Putnam, 1975; Burge, 1979; Devitt and Sterelny, 1989).

   It is difficult to categorize Jerry Fodor's various views in the philosophy of mind in these terms. Perhaps it would be accurate to say that he has become more inclined towards externalism as he has moved further from MIT. Fodor has defended a semantic analog of Chomsky's nativism about grammar, and has long stressed the importance of inner representational mechanisms (1975, 1981). His "methodological solipsism" (1980) is also strongly internalist, and positively opposed to the project of theorizing about organism/environment relations, as seen in what he calls "naturalistic" psychological programs. But in more recent years Fodor has become more interested in links between thinker and environment, in the course of developing a naturalistic theory of meaning (1987, 1990).

7 For an application of this distinction in a widely used textbook, see Krebs and Davies (1987). For a critical discussion see Francis (1990).

8 See Maynard Smith et al. (1985). The language of "constraint" can also be the language of moderate externalism, in discussions of the neutral theory of molecular evolution, for example. For philosophy of science, see McMullin (1988).

9 See also Eshel and Feldman (1982) for an explicit suggestion that the adaptationist concept of an "evolutionary stable strategy" (ESS) be replaced or supplemented by a concept based on population genetics, "evolutionary genetic stability" (EGS).

10 Here I focus on the view in Lamarck's most famous work, the *Zoological Philosophy* of 1809. See also Burkhardt (1977) for more detail on these mechanisms.

11 "More complex" organisms here include insects (Lamarck, 1809, pp. 346–7).

12 Another example of an internalist approach to biology is 19th century German "*Naturphilosophie.*" For discussion of this tradition see Coleman (1971), Bowler (1989), Chapter 4, and Lenoir (1982). I will not discuss the German idealist tradition in science and philosophy here.

13 If a rationalist in epistemology explains the operation of mind in terms of internal principles, but then claims that these internal principles are given to us by God, is this also externalism at one remove? I am uncertain how to handle this type of case. This is certainly a less pure form of internalism than the idea that mind has intrinsic properties of order or principles of development, which have no external source whatsoever. The theological view has important differences from the evolutionary approach, however. The evolutionary explanation of internal mental structure is an

explanation in terms of the sustained impact of the same type of natural environment that thought deals with in the present. The theological view explains inner structure in terms of the one-time action of something quite different from the everyday environment. A detailed discussion of internalist views of the mind would be required to do justice to this topic.

14 See Shapin (1982) for a good survey of sociology of science, and for discussion of the relations between more internal micro-sociological factors (such as distributions of skills) and macro-sociological factors (such as class allegiance). A clear and detailed example of externalist explanation in sociology of science is provided by MacKenzie (1981).

Note also that the work I call "externalist" here need not make use of a "coercive" or deterministic pattern of explanation, as Shapin notes. Here as elsewhere, not all externalism is extreme externalism; the scientist is not a *tabula rasa* with respect to the influence of the political, any more than he or she is a *tabula rasa* with respect to data. The external need not exert its influence via "a sort of mugging" (Shapin, 1982, p. 198). MacKenzie, for example, explicitly opposes a deterministic view, and says that his aim is to "identify 'tendencies' of thought that express the influence of the social situation of the group" (1981, p. 6).

15 An interesting work from the present point of view is Peter Bowler's *Evolution: The History of an Idea* (1989). When Bowler is classifying evolutionary mechanisms he makes use of the distinction between internalist and externalist mechanisms the same way I do, citing Gould (1977). But when he comes to discuss different approaches to the history of science he uses the terms "internal" and "external" in the way associated with the debates between empiricism and the Edinburgh school in sociology of science – experiments are an "internal" factor. So it is difficult in Bowler's framework to recognize the analogy between organisms adapting to environmental patterns and scientists adapting their beliefs to patterns in data.

16 In the social sciences the label "structuralism" is often associated with forms of internalism and "functionalism" with externalism, though this is far from a perfect correlation.

17 Chomsky recognizes that his nativism opposes him to certain traditional "progressive" views, which assert that as humans have no intrinsic nature we need not accept inequality as natural (1988, pp. 163–4). Chomsky is then opposed to some enlightenment positions on the indefinite improvability of all individuals, views associated in particular with Helvétius (1773), although not with Diderot (1774). But Chomsky's scientific aims are not anti-enlightenment.

18 There is also a more externalist side to artificial life research – work on "animats" and classifier systems, for example. See Godfrey-Smith (1994c).

19 Various people have sought to give externalist views for Darwinism having an English home – externalist explanations for the popularity of an externalist theory in a certain environment. Some point to the prevalence of individualism in England, and its embrace of competitive capitalism. Nietzsche perhaps gives a more directly

externalist account: "Over the whole of English Darwinism there hovers something of the suffocating air of over-crowded England..." (from *The Gay Science*, quoted in Himmelfarb, 1959, p. 396).

20 See Piaget (1971) for more on this national effect. Note however that while Piaget claims that the French are attracted on the whole to the internalism of "mutationism" in biology, Piaget includes Darwinism in this category (p. 104). I regard this as an inaccurate view of Darwinism in general – it is odd to call behavioral ecology, a prominent Darwinian subfield which pays little attention to internal factors, "internalist." It is also unusual to see the French as great enthusiasts for Darwin (see Hull, 1988, p. 47), though there are exceptions to this.

21 If we are looking for spatial structure to match intellectual, what are we to make of the fact that the department of linguistics and philosophy at MIT, home of Chomsky and Kuhn, has housed some of the most important recent internalists?

22 See Peel (1971, pp. 144–46) for a sketch of the background, as it relates to Spencer and other 19th century English figures. See also Bowler (1989), Chapter 4.

23 See Heilbroner (1992) for this quote (p. 116) and for a good outline of Owen's life and ideas.

24 This can also be the basis for an argument for scientific realism (Putnam, 1978; Boyd, 1983). Boyd regards himself as replying both to constructivist views and also to strong forms of empiricism which are opposed to realism.

25 The entire passage is interesting.

> Many people, directly and indirectly, have told me that the neutral theory is not biologically important, since neutral genes are by definition not concerned with adaptation. The term 'evolutionary noise' has often been used to describe the role of neutral alleles in evolution, with such a contention in mind. I believe this is a too narrow view. First, what is important in science is to find the truth, so the neutral theory should be of value if it is valid as a scientific hypothesis. Second, even if the neutral alleles are functionally equivalent or nearly so under a prevailing set of environmental conditions of a species, it is possible that some of them, when a new environmental condition is imposed, will become selected. (1983, p. 325)

> The first response is a plea for the importance of the neutral theory as a piece of science which treats a specific phenomenon that is important whether or not it relates to the "big issues" about adaptive evolution. The second response claims that neutral alleles may be important in adaptive evolution after all. On this second point, see also Kimura (1991).

26 I was alerted to this point by Douglas Newdick.

27 Dennett acknowledges the influence of Simon's *The Sciences of the Artificial* in this 1975 paper. He also notes that Simon also uses the term "inner environment" but not in the way Dennett uses it. In Simon "inner environment" just means inner structure, of any kind.

28 In iterated prisoner's dilemma contests, tit-for-tat can invade a social environment dominated by defectors, but only if there are other tit-for-tat players or other cooperative players, and some "clumping" or correlation between the behaviors of contestants. A single tit-for-tat player cannot prosper if everyone else is defecting. Tit-for-tat is a more complex strategy than "always defect" and "always cooperate," and it does well in certain complex social environments (see Axelrod and Hamilton, 1981; Kitcher, 1993; Sober, 1993).

29 Latour also distances himself explicitly from a view in which science is carried along by its own "internal thrust," but, as I understand him, here he is using "internal" in the usual accepted sense, in which experiments and facts about nature are seen as internal to science (1987, p. 175).

30 Some presentations of this idea appeal to the specific statistical sense of "interaction," which is contrasted with an additive relation between factors (Piaget, 1971; Oyama, 1985). Piaget (1971, pp. 99–100) also claims that the internalist and externalist options make use of one, "linear" type of causation and it is the mark of an interaction-based view that it uses a more complex "cybernetic" concept of cause. So the interaction-based view is not a mere intermediate between untenable extremes, but rather it "oversteps them dialectically" (p. 102). Piaget acknowledges the influence of Waddington on his biological thought.

It might be possible to understand Kauffman's biological views in the same type of way (1993), but I read Kauffman as advocating a role for *both* internal and external, rather than trying to discard the distinction.

31 Oyama does say that "the mutual selectivity of stimulus and system applies to causal systems of all sorts, and illustrates the impossibility of distinguishing definitively between internal and external control..." (1985, p. 59). This appears to me to be an over-strong general assertion of the impossibility of externalist or internalist programs of explanation. See also Oyama (1992).

32 The view advocated here is related to, but not identical to, the view proposed as a "test of adaptationism" in Orzack and Sober (1994). The chief difference is that Orzack and Sober envisage a comparison between a purely selectionist model and a more complicated pluralist model. If the selectionist model is as good as the richer one, then, they claim, a form of adaptationism is vindicated. The scenario I envisage, on the other hand, involves a comparison between externalist and internalist (or pluralist) models of comparable sophistication.

33 A strong version of the idea that environmental complexity is sufficient for organic complexity, of a certain kind, is sometimes associated with Ross Ashby (1963), and his "law of requisite variety." This law is expressed by John Staddon as follows: if an organism is to maintain its internal organization constant in face of external perturbation, the variety of its output must equal the variety of its input (Staddon, 1983, p. 89). However, I am not sure if Ashby really intended the law in this strong form. For an ambitious (and in many ways Spencerian) attempt to link environmental uncertainty and organic variability, see Conrad (1983).

65

3

Spencer's Version

3.1 Spencer's place

Herbert Spencer (1820–1903) occupies an unusual place in history. He was unquestionably one of the most prominent intellectual figures in the English-speaking world during the Victorian period. He was largely self-taught, and not affiliated with any university or research institution. He published on biology, psychology, philosophy, sociology, ethics and political science. Spencer is often associated with the label "social Darwinism," although this is a questionable term for someone more inclined towards (what are now called) Lamarckian mechanisms for change.[1] He did advocate an extreme *laissez-faire* position in politics and economics. Spencer was respected, with qualifications, by Darwin and was a close friend of T. H. Huxley. Once when Spencer was in financial straits, J. S. Mill offered to finance some of his writings personally. William James was captured by his views early but later revolted. Spencer was active and influential during the entire second half of the nineteenth century, but soon after the turn of the century his reputation fell like a stone, never to recover.[2]

Spencer is, as far as I know, the first person to use a version of the environmental complexity thesis as the central idea in a theory of the place of mind in nature. This historical claim is made very cautiously. There are difficult questions about the extent to which Spencer's work is continuous with earlier empiricist and associationist thought, and with the work of earlier evolutionists such as Lamarck.[3] Many historians consider Spencer a very unoriginal thinker.

The issue of priority is not my main concern in any case. The aim of this chapter and the next is to outline and compare two very different versions of the environmental complexity thesis, in the hope of gaining a clear view of the resources of the thesis, and also casting some light on the relation between pragmatist and naturalistic outlooks in philosophy.[4]

Spencer constructed a comprehensive and intricate system of the world. This system is probably not internally consistent. In particular, it is not always easy to make the view of mind that Spencer developed in his biological and psychological writings cohere with his official epistemology and metaphysics. My aim here is to focus on the biological and psychological writings. The project is understanding Spencer's view of mind and nature, not the whole Spencerian system of the world. So I will briefly outline Spencer's overall picture, and then isolate the parts that are most relevant to the project of this book.[5]

The idea of evolutionary change was central to Spencer's work. He was the person who originally popularized the term "evolution," and he made use of evolutionary ideas in the broadest and most ambitious ways. He claimed in his *First Principles* to have isolated a general "law of evolution" which applies to the evolution of solar systems, planets, species, individuals, cultural artifacts and human social organizations. This law asserts that there is a universal trend of change from a state of "indefinite, incoherent homogeneity" towards a state of "definite, coherent, heterogeneity" (1872, p. 396). Roughly, this meant that there is a tendency for change from a state where a system has little differentiation of parts, little concentration of matter, and everything is much the same in structure, towards a state in which there is a variety of clearly distinguishable parts, and where the individual parts are densely structured and different one from another. Von Baer, an embryologist, had earlier claimed that the direction of change in individual biological development was towards differentiation. Spencer generalized this idea massively, extending it into every possible domain.[6] Stars cluster into a variety of galactic structures and forms; the earth's surface changes from an undifferentiated soup into a complex of mountains, oceans, deserts and forests; a fertilized egg develops into a cat with bones, blood and fur, and economic systems show increasing specialization of roles and division of labor.

Thus the basic form of change Spencer saw in the world appears to have a "negentropic" quality. It tends towards increased differentiation and organization. This is not to say that Spencer denied outright the global tendency described in the second law of thermodynamics, which was being formulated and discussed around the same time. This law asserts that, in isolated systems, there is a statistical tendency towards a loss of local inequalities in the distribution of heat-energy of the sort that can apparently be made to do useful work. In Spencer's day this law was often interpreted in a pessimistic way, and once alerted to the apparent dire consequences of the second law, Spencer struggled with it. He accepted that the universe as a whole was running down, describing this in his preferred language of "equilibration," or movement towards equilibrium. But though the eventual fate of all forms of living organization is "omnipresent death" through the dissipation of energy (1872, p. 514), as long as the processes of organic and social evolution have the required energetic resources they will tend towards increased organization and differentiation, and the establishment of a state of "moving equilibrium" between external demands and internal responses. This process "can end only in the establishment of the greatest perfection and the most complete happiness" (1872, p. 517).[7]

Spencer claimed that his law of evolution had empirical support, but was also derivable from the most basic principles governing matter and energy. It is here that we reach some of the most notorious parts of his work. The principle Spencer relied on most heavily is a conservation law called "the persistence of force," and interactions between "forces" are his basic explanatory tools. The inferences about forces in the *First Principles*, which Spencer saw as the keystone of his work, comprise an *a priori* pseudo-physics which shadows parts of real Newtonian physics but is also crammed with the most brazen armchair science. This work, along with parts of his political and moral thought, is no doubt largely responsible for Spencer's fall from grace and the depth of the oblivion where he dwells today.[8] This does not pose a problem for us, though. The interesting parts of Spencer's thought for our purposes are largely independent of the supposedly foundational parts.[9]

Spencer thought that a system like a galaxy would get more heterogeneous and differentiated under its own steam. However, once we reach the organic realm, the increases seen in each system's complexity are the consequence of *external* forces. Spencer's basic explanatory pattern in the fields of biology and psychology is an almost unalloyed externalism. (We will look at some reasons for the qualifier "almost" in a later section.) Spencer fused empiricism and associationism about the mind with adaptationist biology, to produce a unified picture of organic activity based around the concepts of adaptation and "equilibration" of organism to environment.[10]

Although I present Spencer's system as a flagship for externalism, Spencer has elsewhere been labelled an "internalist" with respect to the explanation of biological complexity (McShea, 1991; Lewin, 1992). I think this is a misleading categorization. Spencer certainly thought that on the global scale complexity will inevitably develop by itself, as a consequence of natural law. But the manifestation of this trend in organic development is not in any relevant sense "internalist." If we look at the total system which includes an organism or a species plus its environment, this system contains all the ingredients needed to produce differentiation and heterogeneity. But the *part* of this larger system which is biological becomes more complex as a consequence of factors external to it. For Spencer, internal properties of organic systems explain why these systems are the *types* of things which respond to their environments so sensitively, but the particular changes that any organic system undergoes are explained in terms of the specificities of its environment.

The extreme nature of Spencer's externalism in this domain was recognized by T. H. Huxley, who jibed in a letter to him that Spencer seemed to think that "by transferring a flour-mill into a forest you could make it into a saw-mill" (quoted in Peel, 1971, p. 146). Spencer was an extreme externalist even by nineteenth century English standards.

3.2 Life and mind

A central part of Spencer's treatment of the organic world is his definition of life and mind. He had a single definition which applied to both; the distinction between them is one of detail and degree. Mind is an advanced mode of living, and on Spencer's view humans are not just smarter than crabs, they are also in a sense more alive than crabs. The properties that make something a living system, and also those that make it intelligent if it is intelligent, are functional or perhaps "systems-theoretic" properties. They concern the organization of the system's internal components and (most importantly) how they are related to conditions in the system's environment. So this is what we would call a functionalist view of life and mind, in the "dry," causal-role sense of function.[11]

For Spencer, a living system is one distinguished by the existence of certain relations between internal processes and patterns in the system's environment. As a first pass at describing these relations we can make use of an intuitive formulation which Spencer gave in *First Principles*:

> All vital actions ... have for their final purpose the balancing of certain outer processes by certain inner processes. There are unceasing external forces which tend to bring the matter of which organic bodies consist, into that state of stable equilibrium displayed by inorganic bodies; there are internal forces by which this tendency is constantly antagonized; and the perpetual changes which constitute Life, may be regarded as incidental to the main-tenance of the antagonism. (1872, p. 82)

This account of life is based upon the fact that living systems act in a way that tends to be self-preserving. Living systems causally affect the relations between themselves and their environments in such a way that the disequilibria between system and environment are preserved.

In the expression of his view above Spencer uses the term "purpose" which is of course suspicious for any naturalist. Spencer too tried to dispense with this term when he was expressing his theory of life more precisely. In my view the term "purpose" can simply be dropped from the passage above, and replaced with something as simple as "tendency," and the central idea is retained.

The idea that self-preservation is basic to life was not new with Spencer, and I will not discuss the deeper origins of this view. Part of the interest in Spencer's account lies in his recognition of the importance of what we would now call properties of *homeostasis*.

> [T]o keep up the temperature at a particular point, the external process of radiation and absorption of heat by the surrounding medium, must be met by a corresponding internal process of combination, whereby more heat may be evolved; to which add, that if from atmospheric changes the loss becomes greater or less, the production must become greater or less. And similarly throughout the organic actions in general. (1872, pp. 82–83; see also pp. 499–500)

The definition of life which summarizes these features, a formula at the heart of Spencer's biological and psychological views, is: "the continuous adjustment of internal relations to external relations" (1855, p. 374; 1872, p. 84).

The pivotal term in this definition is "adjustment." What is an adjustment? Two different themes in Spencer's discussion are intended to explain this. One theme is seen in the examples I have stressed so far: living systems react to environmental events in ways that tend to preserve their physical integrity and organization. When Spencer is developing this theme he anticipates concepts of negative feedback and homeostasis. But Spencer also has another vocabulary for describing the "adjustment" displayed by living systems. He says life and mind are characterized by a relation of *correspondence* between internal and external.

Spencer's discussions of correspondence are not supposed to replace talk of "adjustment" but to explain it; the *ways* in which living systems succeed in preserving the discontinuities between organism and environment involve relations of correspondence and "concord" between inner and outer. Few relations are more philosophically controversial now than correspondence, and this claim of Spencer's was controversial in his time also. James and

Dewey, for example, both reacted specifically to Spencer's claim that correspondence is a basic relation needed to understand life and mind. James' criticism was an internalist one, which we will look at later in this chapter. Dewey saw Spencer's mistake differently:

> If the organism merely repeats in the series of its own self-enclosed acts the order already given from without, death speedily closes its career. Fire for instance consumes tissue; that is the sequence in the external order. Being burned to death is the order of "inner" events which corresponds to this "outer" order. What the organism actually does is to act so as to change its relationship to the environment; and as organisms get more complex and human this change of relationship involves more extensive and enduring changes in the environmental order.

> [A]ll schemes of psycho-physical parallelism, traditional theories of truth as correspondence, etc., are really elaborations of the same sort of assumptions as those made by Spencer: assumptions which first make a division [between organism and environment] where none exists, and then resort to an artifice to restore the connection which has been willfully destroyed. (1929a, p. 283)[12]

Dewey saw Spencer's use of correspondence as both the product of an artificial "dualistic" divide between organism and environment, and as neglecting the role played by organic transformation of the world. The details of Dewey's account will be examined later. It is true that Spencer's official epistemology, his account of the status of his own work and of science in general in *First Principles*, is "dualistic" in just the sense Dewey criticizes. But Spencer's positive account of the nature of living systems has a different character. Here Spencer is probably guilty on Dewey's second count – neglect of organic action on the world – but not on the first. As I read him, the view of organism/environment relations that is most basic to Spencer recognizes rich causal traffic between the living system and the rest of the world. The only "division" between the living system and the world on this account is a set of physical discontinuities which are constantly under threat and which the organism's actions physically maintain. Spencer's talk of correspondence is supposed to be part of an account of how this is done. What distinguishes living systems from inanimate ones, for Spencer, is the fact that with living systems there is "a connection between the [internal] changes undergone, and the preservation of the things that undergo them" (1866, p. 73).

The idea that life involves homeostasis and self-preservation has been developed in a variety of ways in the twentieth century, especially by workers associated with cybernetics and related fields (Ashby, 1963; Piaget, 1971). On this view a living organism is a natural system whose internal processes and actions upon its environment tend to preserve the system's structural organization and physical distinctness from its environment, in response to environmental events which would otherwise tend to destroy this organization and distinctness. Living systems maintain a characteristic pattern of organization in the face of environmental threats and constant turnover in materials.[13]

There are problems associated with taking a view of this type as a complete account of what is distinctive to living systems. Indeed, many writers are skeptical about the possibility of giving anything close to a definition of, or necessary and sufficient conditions for, being a living system. Looking at biology textbooks, especially, gives one the impression that life is a good example of a "cluster concept"; things that are alive have some appropriate but idiosyncratic mix out of a set of important properties. (See, for example, Curtis and Barnes, 1989, which lists seven distinct basic properties. See also Mayr 1982, Chapter 2.) As well as a property of self-maintenance, it is often asserted that reproduction and/or other population-level properties are necessary for life, or perhaps that certain patterns of transformation of energy are required.

The project of this book does not require proposing a general theory of life. Producing self-maintaining responses to environmental events is, I think, one of the basic activities of living systems. But though a theory of life is tied to Spencer's version of the environmental complexity thesis, this is an optional aspect of the thesis.

3.3   Continuities

Both Spencer's and Dewey's views of mind have close relations to their analyses of life. Some disagreements between them about the place of mind in nature can be related to differences in their views of life. This fact is a consequence of something Spencer and Dewey shared: a thesis of *continuity* between life and mind.

Spencer and Dewey saw cognition as something which emerges out of simpler organic forms of organization and commerce with the world. Dewey called this an assumption of "continuity" (1938). We can distinguish several different possible claims of "continuity" between life and mind.

Weak continuity: Anything that has a mind is alive, although not everything that is alive has a mind. Thought is an activity of living systems.

Strong continuity: Life and mind have a common abstract pattern. The functional properties characteristic of mind are enriched versions of the functional properties that are fundamental to life in general.

These are both constitutive principles, principles about what life and mind *are*. There is also a continuity principle with a purely methodological character.

Methodological continuity: Understanding mind requires understanding the role it plays within entire living systems. Cognition should be investigated within a "whole organism" context.

Strong continuity, as I understand it, implies weak continuity. If the pattern of organization characteristic of cognition is a richer version of the pattern of organization characteristic of life in general, then anything which can think has, in large degree, the properties required for life.

I suppose there could be a thesis of abstract similarity between life and mind that did not have this consequence. Life and mind could be held to have the same basic pattern, but where it is possible for a thinking system to miss out on some necessary property for life. In this book, however, strong continuity will be understood in a way which implies weak continuity.

Weak continuity does not imply strong continuity. Not everything which is only done by living systems is itself life-*like*. Suicide is something that (construed literally) can only be an activity of a living system. But the pattern of suicide does not resemble the abstract pattern of life.

It is a consequence of weak continuity that artificial life must precede artificial intelligence (or in the limit, be simultaneous with it). So in some of the functionalist, computer-based views of mind seen in the 1960's and 1970's we find clear examples of views which deny the weak continuity thesis. There was not a concerted attempt, within these discussions of how thought could exist in computers, to demonstrate that computers of this type would be alive.

Spencer, who had a single definition of life and mind, is a clear example of someone who held the strong continuity thesis.[14] Dewey had a complicated view which will be examined in Chapter 4. I understand him to assert weak continuity only.

Strong continuity is a very tendentious idea, but if it was true it would have a range of important consequences for philosophy of mind. Recent theories of how the unusual properties of mind (such as intentionality and conscious experience) could exist in a physical world have usually been theories that try

73

to establish a direct link between the mental and the physical. I suppose many materialists might accept that, in principle, life exists as a sort of intermediate level between physical and mental. That is, we could in principle explain life in physical terms and then explain mind in terms of life. But life is perhaps thought by modern materialists not to pose difficult philosophical problems, in the way that mind does, so it is assumed that explaining life in terms of the physical would not take us much distance in explaining mind.[15]

If strong continuity is taken seriously, there is an alternative way to proceed. If strong continuity is true, it may be that explaining life in terms of the physical is the hard task, and once that is done, explaining how certain living systems also have mental states is comparatively easy. That is, on this view at least some of the hard problems concern the relation of life to the physical, rather than the relation of mind to the living. Materialists might then try giving a theory of life, upon which a theory of mind can more easily be built. Spencer's method of analysis exemplifies this alternative approach.

This is an interesting possibility for materialism, but strong continuity must be treated cautiously. Against strong continuity it can be argued that cognition is a local organic adaptation like any other. Cognition is an evolved trait, like swimming, photosynthesis, parasitism and camouflage. Why should this particular form of adaptation be singled out as having some important link to life in general? Is this view a product of "zoocentric" biology, in which plants are taken less seriously than animals?

My own position is that strong continuity is interesting but probably too strong to be true.

Weak continuity is more plausible, but still controversial. There is a variety of arguments that bear on it. For example, if it is accepted that having a mind requires having a set of preferences, goals or an "agenda" of some sort, and it is also accepted that these cannot exist without life, then we have an argument for weak continuity. But both of these premises may be contested. My conjecture is that weak continuity is true, but this claim will not be defended here.

The more important question for us here is whether any of these theses are *required* by the environmental complexity thesis. The answer is that neither of the constitutive theses are required. Spencer and Dewey develop their views of mind with the aid of constitutive theses of continuity, but these are optional with respect to the environmental complexity thesis itself.

In this respect cognition could be like swimming. Swimming has a *raison d'etre* which involves specific environmental properties: swimming is a way of getting around in a liquid environment. In addition to this teleonomic connection, swimming also has an instrumental link to this environmental property. But this can be true even if swimming is not the exclusive property of living systems – if inanimate machines can literally swim. Still less does the biological usefulness of swimming in dealing with certain types of environment imply that the pattern of swimming resembles the pattern of life.

The environmental complexity thesis does itself express a version of *methodological* continuity. It describes a link between cognition and the activities of whole organic systems – dealing with complex environments. The environmental complexity thesis aims at an understanding of the place of mind in nature via placing it in a whole-organism context.

Though these claims of constitutive continuity are optional, they are important in understanding Spencer and Dewey. In both writers, the basic nature of the organism-environment relationship is established by claims made about life in general. The specific forms of involvement with environment found in cognition are based upon general principles about the nature of living systems. For Spencer the most important property of living activity is that it constitutes a set of self-maintaining *responses* to environmental conditions which threaten to disturb the "moving equilibrium" characteristic of life. Organic action in general is conceived as *re*-action, and cognition is viewed the same way. To be a living system at all an organism must be able to respond to some level of complexity in its environment. But environments that are comparatively simple do not demand or generate very complex organic forms. Spencer cites the sea as an example (1866, p. 85). Cognition, for Spencer, is a more complex form of organic activity produced only in more complex environments, and it has its *raison d'etre* in the relation between this environmental complexity and the self-preserving efforts of living systems.

Spencer also gives an account of the specific environmental properties that generate complex, intelligent modes of organic response (1855, §§ 127–157). Especially complex organic mechanisms are found in situations where the environmental states that matter to the organism are (i) highly changeable, (ii) spatially or temporally distal, (iii) compoundable, (iv) hard to discriminate, and (v) abstract, involving superficially heterogeneous classes of events. Spencer also thought that a transition from (what we would call) parallel processing to serial processing is important in the emergence of intelligence from more basic forms of life (1855, §§ 168–69).

I will summarize the basic features of Spencer's view before we move on.

(i) Life is understood in terms of a set of relations between organism and environment. Living systems respond to environmental patterns and events in ways that preserve the organization of the living system.

(ii) These adaptive responses are made possible by relations of correspondence between internal and external processes.

(iii) Complexity in environment and in organism is understood primarily as heterogeneity.

(iv) Strong continuity between life and mind: cognition is understood as a specific manifestation of the fundamental reactive properties of living systems.

(v) Cognition is a product of, and a means for dealing with, especially complex environments.

3.4   Homeostasis and cognition

I have said that Spencer's account of living organization anticipates concepts of homeostasis and negative feedback. Homeostatic mechanisms maintain stasis in certain important properties of organic systems, such as temperature, but they do this by means of variation in other organic properties. Homeostasis is often viewed as a very basic property of living systems. In addition, it has seemed plausible to several people to link cognition or intelligence to life in general *via* the concept of homeostasis. Cognition has been viewed as a highly developed homeostatic device (Lewontin, 1957; Ashby, 1963, p. 196). This is one way to develop the environmental complexity thesis in more detail, a way very much in the spirit of Spencer's own discussion.

There is also another reason to look closely here at the concept of homeostasis, and this has to do with the more general project of understanding externalist views of organic systems. Chapter 2 discussed a category of externalist explanation which I called "c-externalist": explanations of internal complexity in terms of environmental complexity. Are explanations of properties of living systems in terms of homeostasis examples of c-externalist explanations? This, it turns out, is a complicated question.

The concept of homeostasis has been important in physiology since its introduction by Walter Cannon (1932).[16] When Cannon introduced the concept he did so by means of a puzzle: organisms are composed of materials which are highly sensitive and unstable. Yet they succeed in holding certain internal properties constant over wide ranges of environmental conditions. Cannon's concept of homeostasis was intended in the first instance to apply just to this condition of stasis in certain organic variables – "a condition which may vary, but which is relatively constant." It meant roughly the same thing as equilibrium in these properties (1932, pp. 19–23). Given the instability of organic materials however, stasis typically has to be explained in terms of the variable actions of other parts of the mechanism. To keep temperature constant a human engages in a wide variety of different activities in different conditions.

This poses a question which is instructive: is an explanation of an organic activity in terms of its homeostatic properties a c-externalist explanation? Homeostasis involves stasis in one set of internal properties and variation in another. When an organism deals with a changing environment by means of a homeostatic device, a c-externalist can claim that the device is one which meets environmental complexity with organic complexity. But there is also a sense in which the organism is meeting environmental complexity with organic homogeneity; an opponent might claim this is the *opposite* of the c-externalist's preferred relation between the internal and external. On this latter view the changeable properties simply effect a buffer that makes the all-important organic stability possible. So is homeostasis really a c-externalist phenomenon or not?

To answer this we first need to say in more detail what homeostasis is. The classical examples of homeostasis, such as temperature regulation in mammals, involve both stability and change. In the 1950's Lewontin argued that because of this fact there is no reason to take stability to be the mark of homeostasis (Lewontin, 1956, 1957). Survival in a large range of environments is the property that counts, and this will involve stability in some organic properties and change in others. Lewontin argued that as a consequence of this we should understand homeostasis simply in terms of survival, not in terms of some property of organic stasis.[17]

It is true that homeostasis involves both stability and change, but the roles that these properties play are not symmetrical. A mechanism is only a homeostatic one if the contribution to survival made by a variable set of properties goes *via* the stasis of another set of properties. Ross Ashby, in a cybernetic discussion of adaptation and homeostasis, expresses this point. He asks: given the fact that homeostasis involves both constancy and change, which is primary?

> The point of view taken here is that the constancy of the essential variables is fundamentally important and that the activity of the other variables is important only in so far as it contributes to this end. (1960, p. 67)

Suppose O_1 and O_2 are different organic variables or sets of organic variables. A homeostatic pattern of response is one for which the following conditions are true: Survival is improved by stasis in O_1, no matter what O_2 does. It is not the case that survival is improved by variation in O_2, no matter what O_1 does. But O_2 does affect survival, as variation in O_2 is positively correlated with stasis in O_1.

For example, dogs pant to lose heat – this is a specialized homeostatic mechanism. Survival is helped by a stability in body temperature (O_1) whether or not the organism varies its behavior with respect to panting (O_2). Panting only helps the organism insofar as it helps keep the body temperature constant.[18]

This definition is not teleonomic; it is possible for something to be a homeostatic device even though this is not its (teleo-) function. Homeostasis as understood here is a dry functional concept.

When Lewontin suggested that homeostasis be understood simply in terms of survival in a wide range of environments, Waddington objected that going into hibernation is not an example of homeostasis, no matter how adaptive it is (1957a, p. 408; see also 1957b). Hibernation is not homeostatic because much of the organism's physiological state is changed. This is a good example. Lewontin's view must indeed recognize hibernation as homeostatic. In my account, hibernation is homeostatic in only a trivial sense. This is a case where the properties O_1 that are kept constant are as trivially linked to survival as they could be; they are almost constitutive of it. The only properties O_1 in virtue of which hibernation (O_2) could be seen as homeostatic are properties such as: respiration occurring *at all*, the body's physical integrity being preserved, and so on. A genuine case of homeostasis is one in which there is a link between O_1 and survival, but O_1 and survival are more distinct properties, so it is not conceptually *necessary* that O_1 contributes to survival. Body temperature and blood sugar level are examples of such properties.

Now we can ask again: is an explanation in terms of homeostatic response c-externalist or not? The question is answered by being more specific about the structure of the explanations involved – especially, being specific about what is being explained. In a context where it is assumed as background that internal stability is required for normal functioning, variation in behavior produced by a homeostatic device is explained in terms of environmental change. This is a c-externalist explanation. But if we then ask *why* this internal stability is required for normal functioning, we are engaged in a different explanatory project. The explanation might mention the need for a precise coordination of chemical actions, where the behavior of these chemicals is very sensitive to surrounding conditions, such as temperature. This is not a c-externalist explanation. It is an explanation of one characteristic of an organic system in terms of other characteristics of its internal make-up. It is an internalist explanation, one which is not in competition with the c-externalist explanation. Homeostasis is a central enough feature of organic life that it has a place within several different kinds of explanatory projects.

So what is to be made of the idea that cognition should be understood as homeostatic, that intelligence is the ultimate homeostatic device? In the narrow sense of homeostasis adopted here, this is not a good general principle. Genuine instances of homeostasis involve the preservation of stasis in organic properties that are nontrivially linked to survival. When some complex activity of an organism contributes to survival this is only genuinely homeostatic if there is some intermediate organic property (O_1) such that the complex activity contributes to the maintenance of stasis in this intermediate property, where this property makes a real contribution to survival. So the intelligent use of fire by humans for the maintenance of bodily warmth is a genuine case of homeostasis by means of cognition. This is a complex organic capacity which makes it possible for humans to maintain their internal body temperature (O_1) constant through a wide variety of environmental conditions, and hence survive in these conditions. But there is no reason to think that everything intelligence does will be nontrivially homeostatic in this way. If good perception and coordination of actions enables an organism to evade a predator or a rock slide, this is not genuinely homeostatic. Cognition here helps the organism to maintain the basic integrity of the organic system. But this is not done by means of the stability of some intermediate organic property such as temperature or blood sugar. Cognition here is like hibernation – it is adaptive, but the explanation for why it is adaptive goes *directly* from organic variation to survival.

So when cognition plays the role described by the environmental complexity thesis, there will be some cases where it functions as a homeostatic device, and others where it does not.

3.5 Spencer's explanatory program

In the next few sections I will describe some of Spencer's views about biological and psychological mechanisms. Then we will look at some internalist objections to these views developed by William James.

Spencer held that organic complexity of every kind was a response to environmental complexity. But did he think that internal factors had no importance at all? In fact he allowed that some explanatory role was played by internal properties of organic systems, but these internal properties had little *specific* explanatory significance. He held that organic systems have, by virtue of their chemical make-up, an extreme sensitivity to impacts and disturbances. They contain a lot of energy stored in chemical bonds which can be released by comparatively small triggering forces, and they are susceptible to dramatic changes in form (1866, §§ 10–16). These factors help explain why organic systems are the types of things which are highly reactive to impacts from outside.

We will look at two aspects of Spencer's externalism here: his view of biological evolution and his associationist psychology.

Spencer was an evolutionist before Darwin's *Origin*, and he saw evolution as environment-driven and adaptive. His views here derive in part from the ideas of Jean-Baptiste Lamarck, and Spencer is often understood now as simply one of Lamarck's followers.[19] But the relation between Lamarck and Spencer on the mechanisms of evolution is more interesting than mere imitation.

In Chapter 2 Lamarck was used as an example of someone who recognized a mixture of internalist and externalist mechanisms for change, where each mechanism is associated with a particular explanandum. Lamarck is remembered for his claim that there can be evolution via the "inheritance of acquired characteristics," the inheritance of properties that organisms acquire during their lifetimes as a consequence of habits they adopt in dealing with their environments. This is often called "Lamarckism" (and in the remainder of this book, when Lamarck himself is not being discussed, we will use that term with its modern meaning). This mechanism can be described in a highly externalist way, and it is this mechanism that Jean Piaget refers to when he says Lamarck's view "might have been specially designed to appeal to the English" (1971, p. 104). But, as outlined earlier, this was not Lamarck's only mechanism for change. Lamarck thought the overall increase in complexity which evolution exhibits is the consequence of other, primarily internalist factors. Complexity is explained internally, and the role of the environment is primarily that of a factor which *disturbs* this trend.

> It is obvious that, if nature had given existence to none but aquatic animals and if all these animals had always lived in the same climate, the same kind of water, the same depth, etc., etc., we should then no doubt have found a regular and even continuous gradation in the organization of these animals. (Lamarck, 1809, p. 69)

Lamarck explained things other than progressive increases in complexity in terms of the environment, and had a primarily internalist account of organic complexity. Spencer had an intrinsic or internal mechanism for directional changes in the environment, and an externalist account of nearly all organic properties, especially complexity.

Spencer's main discussion of the causes of evolution begins with a dismissal of explanations of evolution based upon an "inherent tendency" or "innate proclivity" in living things, as described by people such as Erasmus Darwin, Lamarck and Chambers (Spencer, 1866, pp. 402–408). Spencer finds that earlier writers, if they said anything clear about these "innate proclivities" to evolve, tended to view them as desires. Spencer agrees that desires can probably influence evolution, via changing behavior and hence structure. But where do the desires come from? Where can a desire to make use of a *new* mode of living originate? This leads Spencer to think these evolutionists have started at the wrong place.

So Spencer begins his account of the role of external factors in evolution, and this begins in characteristic style. He first describes a set of very large-scale, rhythmic astronomical events that affect the energy received on earth (1866, § 148). One of these cycles through 21,000 years, another through millions. These cycles are independent, resulting in a "quadruply-compounded rhythm in the incidence of forces on which life primarily depends" (p. 412). Similar phenomena are found in geology; a great variety of types of change to the physical landscape, which work independently and compound upon each other to produce new combinations of conditions. Astronomical changes in concert with geological changes produce meteorological changes, and the chain of effects does not stop there. As a consequence,

> the incident forces to which the organisms of every locality are exposed by atmospheric agencies, are ever passing into unparalleled combinations; and these on the average ever becoming more complex. (1866, p. 416)

This gives Spencer his basic fuel for evolutionary change. But there is a problem with the argument so far. Spencer describes a world with a large stock of processes that have in themselves no directionality, such as astronomical rhythms. These processes cycle independently and hence present organisms with new combinations of conditions. But though organisms will encounter novelty as a consequence of this, they should not meet with a net increase in complexity through time.

This problem suggests a more general challenge that Spencer faces. If all organic complexity comes from environmental complexity, and organisms are eventually going to end up as complex as us, there must be a truly immense store of environmental complexity available somewhere. New combinations of old cycles do not seem enough. This argument had an important proponent in Spencer's time. Louis Agassiz argued against externalist evolutionary views by contrasting the directionality of organic change with the non-directional character of physical changes (see Gould, 1977, p. 11).

Spencer has several additional sources of complexity though. First, as his "law of evolution" suggests, he thought many basic physical changes in the universe are themselves directional, and tend towards increasing complexity. Second, many of the most important influences on organisms are due to the presence and activities of other plants and animals. His conception of nature is holistic: "the plants and animals inhabiting any locality are held together in so entangled a web of relations, that any considerable modification which one species undergoes, acts indirectly on many other species; and eventually changes, in some degree, the circumstances of nearly all the rest" (1866, p. 416). Changes in population size and migrations exert especially strong effects, producing "waves of influence which spread and reverberate and re-reverberate" through the flora and fauna of an area (p. 417). As a consequence of this, "on the average, the organic environments of organisms have been increasing in heterogeneity" (p. 418). This source of environmental complexity is organic, but still external to the organisms whose complexity is to be explained; it is important that externalist views are fully compatible with a strong emphasis on the role of the "biotic" environment, as well as the "abiotic." Further, the complex social environments confronted by humans are the products of human activity (1866, p. 87).

Spencer also has one more card to play:

> Yet a further cause of progressive alteration and complication in the incident forces, exists. All other things continuing the same, every additional faculty by which an organism is brought into relation with external objects, as well as every improvement in such a faculty, becomes a means of subjecting the organism to a greater number and variety of external stimuli, and to new combinations of external stimuli. So that each advance in the complexity of organization, itself becomes an added source of complexity in the incidence of external forces. (1866, p. 418)

This phenomenon has the potential to produce an inexhaustible source of complexity in an organism's experience of its environment. It is a process of positive feedback, in which each additional adaptation made by the organic system acts to generate new complexity in its experience.

Note that this is a feedback process which should apparently work both ways: loss of a sensory mechanism or the development of an additional buffer against the world will lead to a *decline* in experienced environmental complexity. So once an organic system starts down that road there should be a positive feedback process in that direction as well. Spencer does discuss exceptions to the usual trend, the cases of huge trees and hard-shelled tortoises, which meet countless environmental events with the same stolid, static reply. He regards these as unusual cases. In the argument given above though, what we would expect to find in nature is both extremes; there should be hyper-sensitive organisms and also very well-buffered and insensitive ones. The world should contain poets wandering through redwoods, but no beetles or rodents around them.

Spencer can avoid this prediction as a consequence of the *in*organic environmental changes which tend towards increased complexity. Each organism will generally move in the first instance towards increased causal and informational commerce with the world, and the positive feedback process will take it further along that same path.

In any case, given this mechanism Spencer's well of complexity in the experience of organisms is bottomless. But this is only achieved at the expense of a partial concession to internalism. The level of complexity in the "incident forces" is dependent in part on the existing "faculties" of the organism in question. (Note that Spencer does not say that the environment itself has its degree of complexity determined by these faculties.) So internal mechanisms play a role in determining the effects of present inputs. This is a significant concession to internalism, but not a capitulation (see Chapter 5 below).

Though Spencer's evolutionism was originally based on the externalist side of Lamarck, at the publication of the *Origin of Species* Spencer accepted Darwin's mechanism, using his notorious phrase "the survival of the fittest" to describe it (1866, p. 444).[20] Darwin himself held that though natural selection is the major mechanism of evolution, the inheritance of acquired characteristics is probably real and plays some role. Spencer had a similar view, giving a comparatively larger place to the inheritance of acquired characteristics.[21] Later in Spencer's career he defended this pluralistic view against naturalists who were "more Darwinian than Mr. Darwin himself" (1887, p. 29), as they rejected the inheritance of acquired characteristics.

In the twentieth century "Lamarckian" mechanisms for evolution have often been used by internalists, as they leave a place for the organism's active choice in determining its evolutionary future (see Bowler, 1992, for a summary). But evolution by the inheritance of acquired characteristics was often viewed differently in the nineteenth century. Certainly Spencer did not see his acceptance of evolution by the inheritance of useful acquired characteristics as in any way compromising his views on the explanatory primacy of external factors. He incorporated Darwinian and Lamarckian mechanisms into a single integrated view.

Spencer referred to all evolution as "equilibration," and equilibration in biological systems could be either "direct" or "indirect." Direct equilibration covers both individual adaptation, including learning, and also evolution by the inheritance of these acquired characteristics. Evolution by Darwin's mechanism of natural selection was "indirect equilibration" (1866, p. 467). These are both ways in which organic systems are brought into more sophisticated and complete correspondence with external conditions, and brought into this correspondence *because* of the action of external conditions. We will look in more detail at this direct/indirect distinction below. Let us first look at Spencer's mechanisms for within-generation adaptation, and in particular at his view of learning.

Spencer placed great weight on the mechanism of association of ideas which had become central to British thinking about the mind, but he recast associationism within an evolutionary framework. At the center of this part of his work is what he calls a "law of intelligence":

> The law of intelligence, therefore, is, that the strength of the tendency which the antecedent of any psychical change has to be followed by its consequent, is proportionate to the persistency of the union between the external things they symbolize. (1855, p. 510)

More simply, the strength of an association between ideas is proportional to the strength of the association between their objects. Spencer also makes a strong claim about when the "law" will operate.

> [T]he only thing required for the establishment of a new internal relation answering to a new external one, is, that the organism shall be sufficiently advanced to cognize the two terms of such new relation, and that being thus advanced, it shall be placed in circumstances in which it shall experience this new relation. (1855, p. 532)

Spencer accounted for innate reflexes, instincts and habits by arguing that though the individual has not encountered the association in experience before it exists in the mind, the association has been part of the experience of the race as a whole (1855, p. 530). Individually acquired traits of the nervous system can be passed on to the next generation. Spencer also used this mechanism to account for mental associations of such a strength that no individual could receive enough experience to establish them. This enabled Spencer to answer traditional anti-empiricist arguments about knowledge of necessary truths. It also places Spencer as an early proponent of an idea associated with more recent work in "evolutionary epistemology": what is *a priori* for the individual is *a posteriori* for the population or species (Campbell, 1974). Spencer saw himself as using evolution to answer Kantian arguments about the need for prior structure in the organization of experience (1855, pp. 578–583).[22] This move allowed Spencer to acknowledge some truth in the Kantian arguments but hold nonetheless to the hyper-externalist view that "all psychical relations whatever, from the absolutely indissoluble to the fortuitous, are produced by experiences of the corresponding external relations; and are so brought into harmony with them" (1855, p. 530).

Spencer did not think that all useful modifications made during organisms' lifetimes have an intelligent character, and can be treated as associations of ideas. But the law of intelligence exemplifies the basic character of Spencer's within-generation mechanisms. In each case, some property of environmental pattern induces, through interaction with the organism, an organic change which adjusts the organism to this pattern. In particular, additional complexity in environmental patterns induces additional complexity in organic structure.

The "law of intelligence" also illustrates the extreme nature of Spencer's externalism. We can distinguish between externalist explanations for the *material* of thought and externalist explanations for its *pattern*. Locke, for example, insisted that all the materials of thought (simple ideas) were furnished by experience, and he also explained some aspects of the pattern of thought in terms of the pattern of experience. But Locke was also concerned to stress that some properties of complexity of thought need have no basis in experience. The mind can combine simple ideas into complex ones "by its free choice," without consideration of the patterns and connections in nature (1690, Bk. IV, Ch. 4, sec. 5). In contrast, Spencer's explanatory program seeks to link all mental complexity within to complexity without.

We should also note again the distinction between explaining the existence and/or maintenance of cognition as a *whole* in terms of environmental complexity, and explaining *particular* thoughts, or episodes in the cognitive development of particular agents, in terms of specific properties of environmental complexity. The "environmental complexity thesis," as outlined in Chapter 1, is a claim about the former issue. It is a claim about the capacity for cognition itself, rather than specific thoughts. Spencer tried to give basically the same theory to answer both types of question; a single set of principles explains both why thought exists at all, and how thought operates in the day-to-day lives of individuals. But there is no reason why a single theoretical apparatus should do double duty in this way; the mechanisms which generated thought out of the unthinking need not be the same mechanisms that operate in learning and information-processing in individuals.

Spencer's extreme externalism has some interesting consequences. One reason for the modern dislike of Spencer is his adherence to a view of humanity in which Europeans are regarded as more biologically advanced than other ethnic groups. Spencer certainly believed that white European men are the pinnacle of biological development so far, and this is a prominent and unappealing theme in his work. However, it is indicative of the centrality of externalism in Spencer's thinking that this racism is tempered. The superiority of the European is not the kind of thing that can long survive a change of environment:

> In the Australian bush, and in the backwoods of America, the Anglo-Saxon race, in which civilization has developed the higher feelings to a considerable degree, rapidly lapses into comparative barbarism: adopting the moral code, and sometimes the habits, of savages. (1866, p. 190)[23]

3.6 Direct/indirect; instructive/selective

One of Spencer's evolutionary mechanisms was Lamarckian, and his central psychological mechanism was an associationist law. These are mechanisms by which the environment directly impresses its pattern on the organic system; the actions of the environment bring about or elicit an organic change that exhibits a non-arbitrary relation to its environmental cause. In the case of the "law of intelligence," the environment brings about an organic change exactly in its own image.

This type of environmental effect can be contrasted with that seen in a Darwinian mechanism. A Darwinian mechanism is one in which particular organic variants are produced in a random or "undirected" way, and bear no specific relation to the environmental patterns with which the organism is confronted. The role of the environment is to select among the variants produced, retaining some at the expense of others. This is a "generate and test" pattern.

In its modern form, the Darwinian mechanism makes use of internally produced variation and environmental selection. But strictly speaking, the variants need not be internally generated. They can have an environmental cause (and Darwin himself claimed that changes in the "conditions of life" of organisms are necessary for biological variation to occur at all: 1868, vol. 1,

p. 255).[24] The critical feature of a Darwinian mechanism is that the variants produced do not bear a systematic relation to the environmental factors that exert selective pressure on the organism.

A number of more recent writers have made a general distinction between *instructive* and *selective* mechanisms of change and adaptation (Changeaux, 1985; Piatelli-Palmerini, 1989).[25] An instructive mechanism is one in which the environment interacts with the organism in a way that brings about a specific organic change. A selective mechanism is one in which variants are produced by the organism in an undirected way and are then subject to environmental selection. The distinction between Lamarckian and Darwinian evolutionary mechanisms is often used to illustrate the instruction/selection distinction. Immunology is another field in which research has seen a definite shift from an instructive to a selective model. In psychology, classical conditioning has an instructive character, while operant conditioning has a selective character.

This is a useful distinction, especially when applied to simpler externalist mechanisms. The distinction will be used in this book, but it will not be assumed that instruction and selection exhaust the options, with respect to the role of the environment.

Spencer, as I said earlier, made a distinction between "direct" and "indirect" mechanisms of equilibration of organism to environment. It is uncertain how closely this distinction matches the modern selective/instructive distinction, but there is certainly some similarity. Direct equilibration is a process whereby an environmental change or pattern brings about a specific, adapted organic change. Indirect equilibration is a process whereby the adjustment is achieved through "otherwise-produced" organic changes (1866, p. 435). Direct adaptation includes both within-generation adaptation by the individual and also the evolutionary consequences of this change, by means of the inheritance of acquired characteristics. The specific mechanism Spencer discusses under the heading of indirect equilibration is Darwinian evolution by natural selection. This is a process whereby change occurs across generations by the interaction of "otherwise-produced" organic variation and the survival of the fittest.

Spencer saw the connection between associationism about the mind and Lamarckian evolutionary mechanisms. The law of intelligence was linked to Spencer's evolutionary mechanisms in the second edition of the *Psychology* (1871). Darwinian evolution is admitted as the major mechanism for change in the early stages of the evolution of nervous systems, but it becomes less important as the structure becomes more complicated. The inheritance of functionally acquired changes to nervous systems – changes that realize the law of intelligence – is responsible for the later evolutionary stages. It is not strictly accurate to say that Spencer "grouped" Lamarckian evolution and associationism in the category of direct equilibration. For Spencer, direct equilibration is a *single* process, which will be imperfect in a single generation but can be sustained across generations.

In modern selective/instructive taxonomies, two of the paradigm selective mechanisms are Darwinian evolution and trial-and-error learning. Spencer was enthusiastic about Darwinism, but what about trial-and-error learning? It would seem that he was ideally placed to recognize the analogy here, and posit a within-generation mechanism with a selective nature; a mechanism involving undirected production of mental variants, and environmental testing of these variants. In fact Spencer's relation to trial-and-error learning is intriguing.

The first, pre-Darwin edition of Spencer's *Psychology* (1855) does not recognize the possible role of trial-and-error learning. The "law of intelligence" was the basic mechanism for psychological change in this work, and it is not described as a selection process. Then in the second edition of the *Psychology* Spencer retained his discussion of the nature of intelligence as before but added a section in which a physical basis for his psychological principles is given (1871, vol. 1, part 5).[26] Most of this discussion uses direct mechanisms (by Spencer's 1866 definition). But when discussing the origin and development of more complex nervous systems, Spencer introduces a trial-and-error principle for within-generation change. Certain types of nervous connections are established by a process whereby spontaneous, undirected actions produce behavioral success. The process in which the connection is cemented into the system is described in an obscure way. The success produces a "large draught of nervous energy" in the system which somehow flows through and widens the nascent nervous channel responsible for the success. The result is that "what was at first an accidental combination of motions will now be a combination having considerable probability" (1871, p. 545).

So, though Spencer did not change his account of psychological mechanisms, he posited a selective mechanism in the physical realization of these mechanisms. I view this move as *not* compromising the basically instructive character of Spencer's psychological mechanisms. First, the selective mechanism only applies in some cases. More importantly, even when there is intrageneration selection in the nervous system, the law of intelligence retains its instructive character when described at the psychological level: specific new mental structure is produced because of the impact of the structure of the environment. This last claim I recognize as controversial: I assume here that an instructive process can have a selective realization at a lower level. Others may view the introduction of selection as a more wholesale transformation of Spencer's psychological mechanisms.

In any case, though Spencer introduced trial-and-error in this belated and partial way, it was not a very prominent part of his psychological picture.[27] As we will see in the next section, William James saw a major oversight in Spencer's neglect of a pattern of psychological change involving spontaneously produced variants and environmental selection.

In more recent years a number of authors have outlined general frameworks for classifying adaptive mechanisms operating at different levels and time-scales, and have given selective mechanisms pride of place (Popper, 1972; Campbell, 1974; Dennett, 1975).[28] Spencer was engaged in a similar project, but one making use of a different basic mechanism for understanding the role of environment. He sought a general theory of organic structure and behavior in terms of the simpler, one-step model in which environmental patterns directly determine the form of organic response. In this older form of externalist theorizing, organic properties are not seen so much as solutions to environmental problems but rather as bearing the imprint of the environment's patterns. "Adequacy" to the environment can be viewed as a basic relation in both cases, but it is differently conceived and the product of a different process.

It is important to realize how large the gap is between instructive and selective forms of externalist explanation, at least when the instructive mechanisms are simple ones of the Spencerian type. An explanation in terms of an instructive mechanism is not itself an explanation in terms of the utility of an organic property bringing about its retention. The environment's impact brings about new structure in the organism; in principle this could be for good or for ill. The standard examples of instructive mechanisms, and those favored by Spencer, are cases in which the imprint of the environment is also something useful. Spencer's law of intelligence illustrates this. It is a plausible-looking candidate for a directly produced relation between internal and external, but it is also likely to be adaptive. Changes made through "use and disuse" have the same character. Though utility is seen in the paradigm examples, it is strictly irrelevant to the structure of the instructive explanation. The organism may be specifically set up in such a way that direct effects will be adaptive – this would then be a higher-order property of adaptedness. But there is no *a priori* reason why the direct effect of the environment should always bring about something useful (see section 8.4). Spencer tended to try to tackle this head-on; the physical principles through which the direct effect of the environment occurs in his account are such that this direct effect will generally be conducive to the organism's self-preservation. To a modern reader this looks like very wishful thinking.

An explanation expressed strictly in terms of the direct effect of the environment is not a teleonomic explanation in the sense of Chapter 1. So insofar as Spencer sought to show that environmental complexity will bring about intelligence without the operation of selection, his version of the environmental complexity thesis is not a teleonomic claim. To the extent that a selective mechanism is involved, it is a teleonomic claim.

Though it may not be possible to completely assimilate Spencer's view to the teleonomic version of the environmental complexity thesis as expressed in Chapter 1, it is still accurate to ascribe to him *a* version of the thesis. Spencer did see intelligence as adaptive, as functional. He just had some problems with the mechanisms behind this fact.

3.7 James' interests

The first paper in philosophy published by William James was called "Remarks on Spencer's Definition of Mind as Correspondence" (1878). It was a direct attack on Spencer's view of the organic world. James continued this attack in works such as his 1880 paper on "Great Men and Their Environment," and parts of James' *Principles of Psychology* (1890) feature an ongoing to-and-fro with Spencer's ideas (see especially Chapter 28). I will discuss two of James' arguments against Spencer, an argument concerning the idea of correspondence from the 1878 paper, and one from the 1880 paper concerning evolutionary explanation. Both arguments are good illustrations of internalist objections to Spencer's views.

James' basic attack on Spencer in his 1878 paper concerns the role of "subjective interests" in thought. According to James, Spencer's conviction that external relations would force themselves on the thinker and determine the pattern of thought fails to deal with the fact that *which* of the multitude of external patterns a thinker attends to, and how they affect the mind, is determined by the present set of interests, goals and values the subject has.

> No correspondence [between inner and outer] can pass muster till it shows its subservience to these ends. Corresponding itself to no actual outward thing; referring merely to a future which *may* be, but which these interests now say *shall* be; purely ideal, in a word, they judge, dominate, determine all correspondences between the inner and the outer. Which is as much as to say

that *mere* correspondence with the outer world is a notion on which it is wholly impossible to base a definition of mental action. (James, 1878, p.13)

I take it that it is clear how this exemplifies an internalist response to Spencer; a response based upon the explanatory primacy of factors internal to the mind. James continues in a footnote to this passage:

The future of the Mind's development is thus mapped out in advance by the way in which the lines of pleasure and pain run. The interests precede the outer relations noticed. Take the utter absence of response of a dog or a savage to the greater mass of environing relations. How can you alter it unless you previously *awaken an interest – i.e.*, produce a susceptibility to intellectual pleasure in certain modes of cognitive exercise? (1878, p. 13)

James says the interests come *first*, and external facts only play a role determined by this prior filter. According to James, Spencer does give interests a covert role in his account, as all intelligent action is conducive to self-preservation for Spencer. But, James says, self-preservation has no special status; it is one possible interest among many, and it is not true that all intelligent action is subservient to this particular goal.

There is a reply Spencer could make here. Spencer could say to James, as he did to Lamarck: subjective interests and desires matter, but where do *they* come from? They surely do not come from nothing. Perhaps they too come, ultimately, from features of the environment. Then while the organism's interests determine which outer relations will be noticed in the future, these specific interests are themselves a consequence of the action of the environment in the past. This reply would then assert "externalism at one remove," much as Dennett responded to Chomsky (section 2.3 above).

James' argument, as presented so far, is a common one, and Spencer's possible reply appears to herald an extended deadlock. One reason I pick out James' 1878 presentation of this view is the fact that James does not leave Spencer the resources for this reply. He heads off this response by adopting, in contrast to Spencer's extreme externalism, a resolute internalism. I said that Spencer could ask: where do interests come from? Not from nowhere, surely. But James' position in this paper is precisely that they do, in a sense, come from nowhere. James raises the possibility of a physicalist account of the emergence of interests, but says that formulating such a physicalist account is "hopelessly impossible." He concludes:

...I must still contend that the phenomena of subjective "interest," as soon as the animal consciously realizes the latter, appears on the scene as an absolutely new factor, which we can only suppose to be latent thitherto in the physical environment by crediting the physical atoms, etc., each with a consciousness of its own, approving or condemning its motions. (1878, p. 22)

James is not supposing that some mental substance or neural pattern arises literally from nothing. His claim is that the chain of explanation comes to a principled halt when basic forms of interest appear in the world. These are *absolutely new*; there is nothing in the world that, merely repatterned or reclothed, could be a subjective interest. Internal factors have primacy, as interests are "the real *a priori* element in cognition" (p. 13).

This early paper contains internalism of a very high octane. James did not always take so extreme a stand. In later work, especially after he had adopted the label "pragmatism," James located his views closer to the main stream of empiricist epistemology (1898, 1907). Most of James' work can be regarded as fairly moderate, on the issue of externalism, though the emphasis on "interests" was never abandoned.

The second dispute between James and Spencer that I will discuss concerns a more general issue about mechanisms of change. Recall that it is basic to Darwinism to hold that variation is produced in an "undirected" way; the variants produced in a situation bear no systematic relation to what is useful in that situation. Once variants which confer an advantage are produced, they will be retained and proliferate, through the reproductive success of the individuals bearing them. The view of Lamarck, in contrast, holds that the production of variation is related to the needs of the organism.

In twentieth century debate it has often been thought that whereas Darwin's view makes organisms into passive objects of environmental forces, Lamarck's view gives the organism an active role in determining its evolution. As we saw earlier though, in Spencer's hands the Lamarckian view takes on a different character: Spencer's Lamarckian mechanism for evolution is thoroughly environment-driven. It is environment-driven in a way in which Darwin's theory is not, in fact, as Darwinism does not appeal to the environment to explain the occurrence of specific variations.

According to James, a central aspect of Darwin's achievement, and a testament to his "triumphant originality" (1880, p. 221), is his distinction between two types of explanation that can be given for the existence of a trait of an organism. There are firstly explanations of what *produced* a trait, and then there are explanations of the *maintenance* of the trait. If this trait is one the individual was born with, these two explanations refer to two distinct "cycles of causation" which must not be conflated. There is firstly a comparatively mysterious set of physiological causes for the individual being born with that trait as opposed to some other, and then there is a set of environmental factors that explain why this trait is maintained in existence. Similarly, at the level of the population, the physiological mechanisms responsible for the production of variation are to be distinguished from the environmental mechanisms responsible for the retention of certain variants over others. The characteristic mistake of a "pre-Darwinian philosopher," is "the blunder of clumping the two cycles of causation into one" (p. 222).

For James, as the Darwinian picture does not attempt to explain the initial production of variation, but relegates it to idiosyncrasies of physiology, Darwinism does not enslave organism to environment in the way that other evolutionary views, such as Spencer's, do. Darwinism shows how it is possible to leave a definite place for spontaneous individuality in an evolutionary world view. James claims that this message applies also to the evolution of human thought, both in individuals and in societies.[29]

> [T]he new conceptions, emotions, and active tendencies which evolve are originally produced in the shape of random images, fancies, accidental out-births of spontaneous variation in the functional activity of the excessively instable human brain, which the outer environment simply confirms or refutes, adopts or rejects, preserves or destroys, – selects, in short, just as it selects morphological or social variations due to molecular accidents of an analogous sort. (1880, p. 247)

This explains the enthusiasm of a champion of individual spontaneity for Darwinism, a theory often regarded as demeaning creativity. This is not to say that all selective views are internalist. James' use of selection here represents an internalist objection to Spencer because of the explicit stress on spontaneous mental jumps; James' psychological view here is selective and also *saltationist*, rather than strongly gradualist.

Spencer also embraced Darwin's theory when it appeared, and regarded it as no problem for his externalist program. In what way did he see the theory differently from James? The answer is that Spencer was a gradualist. As a consequence, Spencer saw the explanatory structure of Darwinism in a more externalist way than James did.

No version of Darwinism holds that natural selection explains how, against a given genetic background, a particular mutation which confers an advantage arises. That belongs, as James would say, to a different "cycle of causation." But within a gradualist view, natural selection does explain how this particular slight mutation became *accessible* to the population. The mutation became

93

accessible as a consequence of the gradual restructuring of the genetic composition of the population, and natural selection is instrumental in this restructuring. Selection's creative power, with respect to any new mutation, lies in its contribution to the creation of the conditions under which the mutation is accessible.

Although he did not put it in these terms, Spencer recognized the creative aspect of natural selection, and saw this as Darwin's distinctive insight. To Darwin "we owe the discovery that natural selection is capable of *producing* fitness between organisms and their circumstances" (1866, p. 446). Thus Spencer could embrace both Darwinian and also Lamarckian mechanisms of evolution within his hyper-externalist framework, distinguishing only the "directness" of the equilibration involved in each.

Notes

1　See Peel (1971) and Bowler (1989) for objections to the classification of Spencer as a "social Darwinist." They argue that he was primarily a Lamarckian thinker, and that this carries over into his social thought.

2　How low are Spencer's stocks at present? Here is Ernst Mayr: "It would be quite justifiable to ignore Spencer completely in a history of biological ideas because his positive contributions were nil" (1982, p. 386). Spencer does rate a total of two pages in Mayr's monumental history. He is not discussed in Bertrand Russell's *History of Western Philosophy* (1945). Spencer's life, and his intellectual rise and fall, are well documented in Peel (1971) and Richards (1987). For Mill's offer see Heilbroner (1992, p. 135).

3　See Young (1970) for a discussion of the history both of associationist views and of "functionalist," behavior-oriented approaches to psychology.

4　Ruse (1986) uses Spencer's views, as I do, as a starting point in developing a biological approach to thought and knowledge. We stress different parts of Spencer's system: I stress his externalism while Ruse stresses his progressivism. In addition, at several points our interpretations of Spencer and other thinkers (such as Hume) differ substantially. Ruse also discusses a Darwinian approach to ethics.

5　The works of Spencer's that are most relevant to us are his *Principles of Biology* (1866) and *Principles of Psychology* (first edition 1855, second edition 1870). Sometimes *First Principles* will be relevant. The first edition of the *Principles of Psychology* has particular historical interest as it was published before Darwin's *Origin of Species* (1859). This 1855 edition of *Psychology* is now a rare book, and libraries tend to have later editions. Some of the discussions of life and mind we will look at are not found intact in later editions of *Principles of Psychology*, but are shifted to *Principles of Biology*. References to *Principles of Psychology* will be either to the first edition of 1855 or the American second edition of 1871. The *Principles of*

Biology was originally published in parts, 1864–6. References are to the American first edition of 1866. References to *First Principles* are to the American second edition of 1872. Both pages and section numbers (§) will be used, as appropriate. Section numbers are not always constant between editions, though Spencer often gives a key to changes in section number. The main secondary sources I have used on Spencer are Peel (1971), Young (1970), Kennedy (1978), Boakes (1984) and Richards (1987).

6 For Von Baer see Lenoir (1982) and Coleman (1971). Spencer acknowledged the debt: "the law of organic development formulated by von Baer, is the law of all development" (1857, p. 35; see also 1904, Vol. 2, p. 194).

7 See Kennedy (1978), p. 43, for a discussion of Spencer's problems with the second law. I am indebted to Yair Guttmann and Paul Pietranico for helping me with these issues.

Spencer also held out the possibility that after the present organization in the universe had run down into omnipresent death, there might be a restarting of the evolutionary processes that tend towards heterogeneity and order. That is, there could be a global cycling of evolution and dissolution (1872, §§181–83).

8 William James expressed a common reaction to these parts of Spencer's work with characteristic succinctness:

> [T]he spencerian 'philosophy' of social and intellectual progress is an obsolete anachronism, reverting to a pre-darwinian type of thought, just as the spencerian philosophy of 'Force,' effacing all previous distinctions between actual and potential energy, momentum, work, force, mass, etc., which physicists have with so much agony achieved, carries us back to a pre-galilean age. (1880, p. 254)

For the development of James' overall view of Spencer see the discussion in Perry (1935), Chapter 28. James paraphrased Spencer's basic law of evolution, discussed in the text, as follows: "Evolution is a change from a no-howish untalkaboutable all-alikeness to a somehowish and in general talkaboutable not-all-alikeness by continual stickingtogetherations and somethingelseifications" (quoted in Perry, 1935, p. 482). A browse through the footnotes of James (1890) also tends to turn up some amusingly peppery remarks. But see also the grudgingly favorable assessment at the end of the first chapter of James' *Pragmatism* (1907).

9 According to Spencer's official "First Principles" epistemology, knowledge is of appearances or phenomena only. But there is an Unknowable something that lies behind and is in some way responsible for these phenomena (1872, Part 1). These epistemological scruples play little role in most of Spencer's concrete discussions of biological and psychological issues.

10 Before 1855 or so, environmental factors were less central to Spencer's writings on life and mind. (Contrast the definitions of life in Spencer, 1852b and 1855, for example.) I focus throughout this chapter on the views Spencer developed in his

central works on these topics, the *Principles of Psychology* and *Principles of Biology*.

I also will not discuss Spencer's social thought, though there are some interesting issues concerning the fit between Spencer's preferred forms of explanation and a social subject-matter. Peel (1971) is a classic and very engaging exposition, with a good discussion of Spencer's attempt to view the category of "adaptation" as basic in the social sciences.

11 This also appears to be a materialist position. In fact Spencer rejected that term, for a number of reasons (1871, pp. 616–627). Some of these are close to some modern functionalist reasons: Spencer identifies mind with *motion*, not with the matter itself, with the music coming from the piano rather than the piano itself. More importantly for Spencer himself though, he held that what lies behind experience is an Unknowable, forever beyond comprehension. Our concept of the "physical" is as much an interpretation of experience as our concept of the "mental" is. Though materialism is rejected as a metaphysical position, Spencer's scientific picture is materialist in its explanation of mental properties in terms of the physical. Spencer sought to understand thought as a functional property of certain complex physical systems.

12 Spencer actually discusses this case involving burns himself (1855, pp. 510–11; or 1871, p. 409), and distinguishes it from genuine cases of correspondence. The organism suffers because of its failure to internalize the fact that fire is dangerous.

13 There is also a more internalist approach to life that is closely related. In this view responses to a structured environment are not the focus, but activities of "self-production" by the organic system (Maturana and Varela, 1980).

14 A more recent example of strong continuity is seen in Maturana: "living systems are cognitive systems, and living is a process of cognition" (Maturana, 1970, p. 13).

15 Older philosophical discussions of mind and materialism do sometimes discuss life explicitly. Broad (1925) is an example.

16 Cannon gives several nineteenth century references which he takes as stressing the phenomena of stability which interest him. He does not mention Spencer. Here is a quote from a Belgian physiologist, Léon Fredericq:

> The living being is an agency of such a sort that each disturbing influence induces by itself the calling forth of compensatory activity to neutralize or repair the disturbance. The higher in the scale of living beings, the more numerous, the more complicated and the more perfect do these regulatory agencies become. They tend to free the organism completely from the unfavorable influences and changes occuring in the environment. (1885, quoted in Cannon, 1932, p. 21)

Lamarck also mentions these factors in his account of life. He did not make them as central to his account as Spencer did (Lamarck, 1809, p. 193).

96

17 This issue was raised by Lewontin during some discussions we will examine more closely in Chapter 9 (Lewontin, 1956, 1957, 1958). It was claimed at this time that individuals of a certain genetic type (heterozygotes) are more homeostatic than others. The concept of homeostasis was associated by some of these writers with *stasis in the face of environmental change*. Constancy *per se* was seen as homeostatic, if maintained despite environmental perturbation. Lewontin saw this as a mistake: not any organic constancy is a sign of homeostasis.

18 This formulation of homeostasis could also be expressed more precisely in probabilistic terms.

Homeostasis:

(i) $Pr(\text{Survival}|\text{Stasis in } O_1) > Pr(\text{Survival}|\text{Variation in } O_1)$

(ii) $Pr(\text{Survival}|\text{Stasis in } O_1 \text{ \& Stasis in } O_2) \geq$
     $Pr(\text{Survival}|\text{Stasis in } O_1 \text{ \& Variation in } O_2)$

(iii) $Pr(\text{Stasis in } O_1|\text{Variation in } O_2) > Pr(\text{Stasis in } O_1)$

This is close to a case of "screening off" (see Salmon, 1989). If (ii) is an equality, then O_1 "screens off" O_2 with respect to survival. The difference is that I assume that variation in O_2 may have an independent energetic cost – a negative effect on survival, but a cost which is worth paying.

This formula would have to be changed if O_2 had some other positive contribution to make to survival as well. For example, sweating is homeostatic with respect to body temperature, and it is costly in terms of dehydration, but we sweat out some unwanted chemicals also (and I assume here that this is an independent benefit). To capture this we could also condition upon the presence, and on the absence, of this secondary positive effect.

19 Spencer first encountered Lamarck's ideas through the criticism of Lamarck by Lyell (Spencer, 1904, vol. 1, p. 201). For a brief early statement of his evolutionary views, see Spencer (1852a).

20 Spencer had often used the term "fitness" in earlier work to refer to a relation between living systems and their environments (1886, Chapter 2; 1852b). But the "survival of the fittest" made its first appearance here in *Principles of Biology*, as Spencer confirms in a footnote to a revised edition of the work (Spencer, 1898, p. 530; *contra* Peel, 1971, p. 137, who attributes the phrase to Spencer's 1852b paper on Malthus and population growth, the paper which came close to anticipating Darwin and Wallace).

21 Spencer claims empirical evidence for the inheritance of acquired characteristics, but (typically) he also has a more *a priori* argument for it. This argument is based upon a holist view of organisms, and the "persistence of force." If an organism is modified by its circumstances, this modification will affect *every* part of its functioning, including the functioning of its reproductive apparatus. So the

off spring must be affected in some way, or there must be a denial of the persistence of force. Other speculative arguments about equilibration suggest that the offspring should vary in the same direction as the parents, as well (1866, §84).

22 I do not know if Spencer was the first person to raise the possibility of this reply to Kant. Campbell (1974) gives a detailed list of people who have advocated this idea. Spencer is the earliest on that list. For evolutionary epistemology see also Plotkin (1982), Ruse (1986) and Bradie (1994).

Spencer's commitment to innate structure in the individual distanced him from some other empiricists of his day, such as J. S. Mill. Spencer also thought that Mill neglected the epistemological role of what Spencer called "the universal postulate" – the principle that any proposition whose negation is inconceivable must be true (1855, 1871). This looks like a rationalist or intuitionist principle, and Mill criticized all such principles. But in fact Spencer's universal postulate did not violate his empiricism. The postulate has a useful role *because* of the long-term effects of experience on the structure of the mind.

23 See Peel (1971, pp. 144–46) for an interesting discussion of the relations between very strong externalist views, such as Spencer's, and racism.

24 Darwin made a variety of claims at different times about the connection between variation and changes in organism's "conditions of life." These claims differed in strength, and the claim in the text is a strong one. (See also Darwin, 1859, p. 82.) See Winther (in preparation) on these issues.

The first and most detailed example of a Darwinian explanation Spencer gives in his discussion of evolution in *Principles of Biology* involves an environmental cause for variation (1866, p. 446).

25 There is some variation in terminology. Some use "inductive" instead of "instructive" (Changeaux, 1985), but this is too loaded a term here. The distinction between Popperian views and more orthodox empiricist positions in the philosophy of science is sometimes understood in terms of a selective/instructive distinction. But one can accept inductive concepts of justification even within a selective view of the mechanisms of theory choice.

For a different kind of selection-based discussion of scientific change, see Hull (1988).

26 Spencer says this section had been planned for the first edition but not included.

This section contains some interesting arguments about the adaptive value of certain basic configurations of neural wiring. As in Braitenberg's *Vehicles* (1984), Spencer discusses the value of connections between brain and muscles which are crossed over, so the left side of the brain controls the right side of the body (1871, pp. 535–36).

27 It has been claimed by Boakes (1984, p. 13) that Spencer took the idea of a trial-and-error mechanism straight from Alexander Bain's work, without acknowledgement. Bain had given what Boakes and others regard as the first psychological discussions

of trial-and-error learning in 1855 and 1859, between the first two editions of Spencer's *Psychology* (see Bain, 1859).

In the second edition of Spencer's *Psychology* he also included a new discussion of pleasure and pain (Part 2, Chapter 9). Pleasures are incentives to life-supporting acts and pains are deterrents from life-destroying acts (1871, p. 284). Their role is a product of natural selection (p. 281). This discussion of pleasure and pain is not integrated by Spencer into his theory of intelligence, or even explicitly linked to the new discussion of trial-and-error in the nervous system. Spencer had all the ingredients in place in the second edition of the *Psychology* to give an explicit discussion of variation and selection on the psychological level, but he did not reorganize the discussion of intelligence in the required way.

28 See Ruse (1986) for criticism of Popper's (1972) version of this idea. Campbell (1974) is a useful survey of the field. Dennett's (1975) version is particularly interesting. See also Amundson (1989) for a critical discussion of these ideas.

29 James was not the first to see the analogy between trial-and-error learning and Darwinian evolution. James may have got the idea from his associate Chauncy Wright (Amundson, personal communication).

4

Dewey's Version

4.1 Meetings and departures

Spencer and *Dewey*? Spencer the scientistic, laissez-faire Victorian who believed in laws of progressive change and explained mind in terms of astronomical rhythms making our lives more complicated... and Dewey the great American liberal, the man who thought ontological guarantees of progress only deflect people from bringing improvement about themselves, and who thought intelligence functions in the *transformation* of environments? Dewey has been befriended by foes of systematic philosophy such as Richard Rorty (1982). Even postmodernists approve of Dewey. For people like that Spencer embodies all the callousness, arrogance and folly of science-worship and system-building.

It is not an obvious combination, but the idea I am calling the environmental complexity thesis lies close to the heart of Dewey's epistemology. Let us have some cards on the table. Here is a passage from Dewey's *Experience and Nature* (1929a) which I take to express a version of the environmental complexity thesis:

> The world must actually be such as to generate ignorance and inquiry; doubt and hypothesis, trial and temporal conclusions; the latter being such that they develop out of existences which while wholly "real" are not as satisfactory, as good, or as significant, as those into which they are eventually reorganized. The ultimate evidence of genuine hazard, contingency, irregularity and indeterminateness in nature is thus found in the occurrence of thinking. (1929a, p. 69)

In Dewey's epistemology an important role is played by a contingent fact about the pattern of nature: the balance which actual environments display between the stable and the unstable, the reliable and the capricious. This contingent fact about the world's composition is part of the reason why inquiry exists. The hazardous quality of the world can be inferred from the existence of thought, much as some properties of a town's climate can be inferred from the town's owning snow-ploughs. Hazard in the world is a necessary condition for the existence of thought.

The environmental complexity thesis is a key part of Dewey's attempt to overcome the separation between mind and nature seen in traditional episte-mology, to show that the scientifically minded philosopher can find a strand of truth in "idealist" philosophies, and to turn philosophical attention away from artificial pseudo-problems towards the concrete issues that humans face. Dewey's work shows how the environmental complexity thesis can operate within a worldview which is in many ways the opposite of Spencer's. Some of the interpretations of Dewey in this book are unorthodox, as they make more of the naturalistic and realist side of Dewey's later work than has been common in recent discussion. But now is a good time to re-evaluate these parts of Dewey's thought. Of Dewey's major works, it is *Experience and Nature*, in particular, that focuses on these ideas in most detail, and this work will be central to my discussion.[1]

This chapter will begin by discussing some points of overlap between Dewey's views and Spencer's. A basic similarity between Spencer and Dewey is their willingness to theorize about the structure of the environments confront-ed by knowers, and the specific role played by environmental complexity.[2] Two other similarities are (i) their partial agreement about life in general and its relation to inquiry, and (ii) their emphasis on concepts of equilibrium and (what we would call) homeostasis in describing inquiry.

Then in the rest of this chapter, and in the next, we will look at divergences between Spencer and Dewey, and at aspects of Dewey's view that will be retained in the version of the environmental complexity thesis endorsed in this book. Three differences will be discussed in this chapter:

(i) In understanding the place of mind in nature Spencer privileges questions about origins, the deep historical roots of things, and the contribution present activities make to a progressive sequence of events originating in the distant past and extending into the future. There is no sharp distinction between teleonomic or historical issues on the one hand, and present tense, instrument issues on the other. Dewey's view of these matters is more complicated, but he recognizes, in effect, the distinction between teleonomic and instrumental, and he does not intend his epistemology to bind the future with the aid of the past.

101

So it is possible to sharply distinguish teleonomic and instrumental issues when discussing Dewey.

(ii) Spencer's preferred explanatory mechanisms make use of the "direct" effect of the environment on organic systems, rather than cycles of variation and selection. Dewey's view of the role of environment is, again, more complex. Dewey's view of the role of environment in its interaction with mind is closer to a selective view than an instructive view. But it is not a paradigm case of a selective view. In addition, Dewey is a far less extreme externalist than Spencer.

(iii) For Dewey, the environments in which cognition has a special role to play are not simply uncertain or complex environments, but environments which contain a *balance* of the unpredictable and the predictable. If an environment is too simple and predictable, there is nothing that needs to be investigated, but if an environment is too chaotically complicated there is nothing which *can* be known.

After discussing these topics, an entire chapter will be set aside to discuss what I regard as the single most important difference between Spencer and Dewey in this domain. Spencer conceives organic action as *re*-action, as a response to external forces which develop in a largely autonomous way and lay down their own law. Dewey's account is based upon two-way connections between organism and environment, and the power of intelligent action to transform the world.

At the heart of Spencer's picture is a network of impersonal, dynamically self-sufficient forces, becoming more and more complex in their manifestations, and impinging upon organic systems in more and more complex ways. This results in increases in organic complexity, and intelligence. At the heart of Dewey's picture is a process in which organic systems act to transform and restructure their environments as a means to overcoming problems. This restructuring constantly gives rise to new problems and new possibilities for intervention and control.

4.2 Dewey on life

We will begin by looking at Dewey's theory of life. Dewey viewed life in general as a property of organization exhibited by some natural systems. Most importantly, the activities of living systems have certain relations to their environment; life is a "transaction extending beyond the spatial limits of the organism" (1938, p. 32). States of organic equilibrium are disturbed, and then, by means of action on the environment, equilibrium is restored. Dewey admits that there are inanimate systems which display a pattern of disequilibrium followed by restoration of equilibrium. What is distinctive of living systems is the fact that they restore equilibrium in a way which acts to maintain the system's organization; the system acts to preserve its integrity in the face of possible disruption and decay.

> Iron as such exhibits characteristics of bias or selective reactions, but it shows no bias in favor of remaining simple iron; it had just as soon, so to speak, become iron-oxide. It shows no tendency in its interactions with water to modify the interaction so that consequences will perpetuate the characteristics of pure iron. If it did, it would have the marks of a living body, and would be called an organism. (1929a, p. 254; see also 1929b, p. 179)

The most basic mark of life for Dewey is self-maintenance. In the previous chapter I isolated this idea within Spencer's treatment of life, and noted that this has become a common view in the twentieth century. It is a functionalist conception of life (in the dry sense of function), a view based on patterns of interaction and causal role, rather than material properties of the living. Whether or not these properties can be made into a complete picture of life, they comprise one of the basic features of living systems.

Dewey also resembles Spencer in the stress he placed on concepts of equilibrium. The idea of equilibrium is not only important in Dewey's view of life, but plays a role also in his epistemology, via his acceptance of a thesis of continuity between life and mind.

This is a point where the historical project of this book must be distinguished sharply from the nonhistorical, positive project. My aim is both to give an accurate picture of a part of Dewey's system and also to isolate some parts of it for actual endorsement. There is a range of claims made by Spencer and Dewey about life which are important in understanding their views, but which are strictly optional with respect to the environmental complexity thesis. Claims made by Spencer and Dewey about equilibrium fall into this category.

Both Spencer and Dewey used the concept of equilibrium in a loose and informal way. Spencer saw living systems as existing in a "moving equilibrium" with their environments, which is to be contrasted with the static equilibrium with environment that dead and inorganic things reach. Dewey also favored this type of language: "living may be regarded as a continual rhythm of disequilibrations and recoveries of equilibrium" (1938, p. 34). For Dewey one of the basic functions of many organic activities, including thought, is aiding in this recovery of equilibrium. States of organic disequilibrium constitute states of need, and thought and other activities are means for meeting these needs, restoring equilibrium.[3]

Once suitably clarified, some of these ideas may be useful, but they are not taken on board here as parts of my preferred version of the environmental complexity thesis.

My grounds for caution here are the same as they were in the discussion of homeostasis in Chapter 3. One way in which people have tried to locate the role of cognition within the activities of living systems is via the idea that cognition is a homeostatic mechanism. But I accept a narrow conception of homeostasis (section 3.4). On this conception it is unlikely that everything cognition does for organisms is homeostatic. For the same reasons it is also unclear that all living and life-preserving activities should be viewed as contributing to the maintenance of properties of equilibrium, unless maintaining some specific equilibrium is exactly the same property as being alive.[4]

So there is a point of partial agreement between Spencer and Dewey on the basic features of life. This view about life is part of a common foundation upon which these two different versions of the environmental complexity thesis are built. Another point of partial agreement concerns their views on the *relevance* of general properties of life to epistemology and the philosophy of mind.

4.3 Dewey on continuity

In the previous chapter I distinguished three theses of "continuity" between life and mind. According to the weak continuity thesis, mind emerges from more basic properties of living organization, and anything which thinks must be alive. According to strong continuity, the general pattern of organization characteristic of mind resembles that of life in general. In addition to these two constitutive theses, there is a methodological continuity thesis, which is independent. The methodological continuity thesis claims that mind should be investigated as a component in whole living systems.

Spencer was an advocate of strong continuity. Consequently his positions about life in general are directly relevant to his view of mind. Dewey was also an enthusiast for "continuity" between life and mind. But his position on this issue is more complicated than Spencer's.[5]

Dewey is certainly committed to weak continuity. He claims that life and mind involve different degrees of a basic property of complexity with respect to interactions: "The distinction between physical, psycho-physical [living], and mental is thus one of levels of increasing complexity and intimacy of interaction among natural events" (1929a, p. 261). Dewey resisted expressing this in the terms of materialism or mechanism; he did not want to say that life is just a complicated characteristic of material systems; the term "material" applies only to natural events when they exhibit certain simple, inorganic patterns of interaction, so it is inappropriate to use that term of nature when it displays a more complex organization.

It is Dewey's position on strong continuity that is complicated. Here he makes two claims. First, the basic pattern of inquiry, or problem-solving by means of thought, does resemble the pattern displayed by the fundamental properties of life. The pattern characteristic of life "foreshadows" the pattern characteristic of inquiry (1938, p. 40). Life and inquiry are both characterized by a sequence involving a disturbance, imbalance or problem encountered by an organic system, and a subsequent organic response that solves the problem or adjusts to the disturbance, via a change made to organism or environment. Life and inquiry are both characterized by equilibrium lost or threatened and then regained or secured. The disequilibria in living organisms induced by environmental events are like "proto-problems" for Dewey, and all living activities which act to regain the organic equilibrium are proto-solutions.

So far these claims are consistent with strong continuity. But another of Dewey's central claims about mind is not. Dewey claims that only an organism which inhabits a social environment and makes use of linguistic communication can literally think (1929a, Chapter 7; 1938, Chapter 3). "Language" is understood in a broad way by Dewey; it is symbolism in general that is required. For Dewey, thought is necessarily symbolic and symbolism is necessarily social. So Dewey is committed to the view that most animals do not think. He accepts this conclusion (1929a, p. 230). More complex nonlinguistic animals can have "feelings," but not representational thought. Though nonlinguistic animals in their life-activities may go through a pattern which resembles or "foreshadows" that of inquiry, it does not involve the deployment of symbols with determinate content, and it is not thought.

I take all views on which public language is necessary for thought to be views which deny strong continuity, unless a very unusual view of the nature of life is taken.

Dewey's claim that symbolism is necessary for genuine intelligence and inquiry is not disconnected from the rest of his views; he is not saying that if a nonsocial animal carries out an apparent piece of problem-solving, all it would take is a clothing of this act in a social context for genuine thought to be the result. Rather, it is characteristic of intelligence to approach problems in a particular way. An environment is approached intelligently when it is approached indirectly – solutions are tried out in thought before they are acted on. Actions can be rehearsed and assessed in symbols without behavioral commitment.[6] For Dewey this cannot be done without symbolism, and symbols with determinate content only exist in a social context. Presumably then, an individual organism with no social connections in its life at all, which engaged in internal processing that looked for all the world like planning based on an assessment of the consequences of various actions, would not really be *representing* to itself particular actions and possible events in the world.

So Dewey's overall view is one of weak continuity only. He does accept that there are similarities of pattern between mind and life in general, but one of the fundamental properties of cognition – its symbolic nature – has no correlate in the basic properties of life.

As outlined in Chapter 3, neither weak nor strong continuity is required by the environmental complexity thesis itself. These continuity claims are strictly optional. In my own view, Dewey's position on continuity is not plausible, as all views that require social interaction for the existence of representational thought are not plausible. This is a difficult issue, as there is no accepted naturalistic theory of what it is for an inner state to have representational content; we will return to some aspects of this question in Chapter 6. But the idea that nonsocial animals do not have thoughts with content is surely a position that we should only accept if we are forced to (although some have willingly taken it on board: Davidson, 1975). There is a variety of theories which try to explain mental representation in a naturalistic way, and which do not make any essential appeal to social interaction or public language (Millikan, 1984; Dretske, 1988). These theories have a range of problems, especially when they are asked to determine the meaning of inner states with the precision reflected in attributions of content between humans. But if the goal is to analyze a simpler, more coarse-grained capacity of internal representation, a capacity which animals can plausibly possess, several of these theories do a good job. Not all representation need have the sophistication and the fine-grained semantic properties seen in human mental and linguistic life.[7]

4.4 Indeterminacy and complexity

Though constitutive claims of continuity between life and mind are optional, these ideas are important in the project of understanding Dewey's view of mind.

Dewey's epistemology is primarily a theory of problem-solving. It is common within pragmatist philosophies to regard cognition as a response to problems encountered in experience. It is distinctive of Dewey in particular to understand these "problems" in terms of specific properties of environments: variability as a property of nature is the source of problems to which cognition is a response. It is this link that makes Dewey's epistemology of problem-solving into a version of what I call the environmental complexity thesis.

Dewey used a range of terms to describe the conditions which prompt inquiry. Thought is a response to the unsettled, the doubtful, the hazardous, the precarious, the indeterminate, the irregular, the uncertain, and so on. Dewey insists that properties like these are real characteristics of environments, not properties imposed on them by thinkers. It is not part of my project to work through the long list of properties which Dewey discusses in this context, and work out which ones can reasonably be seen as properties of environments as opposed to properties of agents themselves (see Thayer, 1952).[8] In this section my aim is to make one specific link between the structural properties of environments and the problems that cognition deals with. What is needed is a link between the concept of environmental complexity which is assumed in this book – heterogeneity – and the properties Dewey uses in his theory of inquiry.

Dewey, in *Logic*, says that inquiry is a response to an "indeterminate situation." A situation is a local episode in an organism's interactions with its environment. Indeterminacy is a property of disturbance, precariousness or disequilibrium in these interactions. It involves uncertainty in what environmental conditions "import and portend in their interaction with the organism" (1938, p. 110). It is important that indeterminacy is a property *recognized* by the agent; thinking that a situation is indeterminate does not make it so.[9]

An indeterminate situation is not the same thing as a complex or heterogeneous environment. But there is a link between them. The link between variability as a physical pattern in environments and indeterminacy as a property of situations is a consequence of the special role that equilibrium plays in Dewey's theory of living organization. As a consequence of the role played by equilibrium in life, environmental heterogeneity tends to give rise to situations that pose problems for organisms, the type of situation that "creates doubt, forces inquiry, exacts choice, and imposes liability for the choice which is made" (1929a, p. 421). An organism acts, qua living system, to maintain a state of equilibrium. Variability and change in the environment impinge on the organic system, and disturb or threaten to disturb this equilibrium. Action which transforms the situation functions to preserve a preferred state.

Environmental variability, unlike environmental stability, is "a call to effort, a challenge to investigation, a potential doom of disaster and death" (1929a, p. 51). In the absence of variability there is the possibility of a fixed, perfectly reliable response; variability is associated with the problematic in a way in which stability is not.

Thus the asymmetric roles played by stability and variability in Dewey's epistemology are linked to the roles played by equilibrium and disturbance in his theory of life. It is when Dewey's epistemology is linked to his general theory of organic systems and their relations to their environments that the place of the environmental complexity thesis in Dewey's epistemology is made clear.[10]

I have presented this link between the properties of life and the properties of problem-solving as an explanatory relation; *because* organic equilibrium has to be maintained, environmental variability generates problems. We should note that there is another way in which Dewey's claims could be interpreted here. The other possible relationship between variability as a threat to life and indeterminacy as the source of epistemic problems is a relationship of mere similarity, without an explanatory connection. That is, Dewey's claims about continuity could be understood to say the following: variability threatens life *and* indeterminacy is what thought deals with, but it is not the case that indeterminacy is problematic *because* variability is threatening to life. The similarity between the pattern of life and the pattern of inquiry would, on this alternative view, be unanalyzed, a similarity worth remarking on but not one with any explanatory arrow. There are difficult issues of interpretation here, but I understand Dewey as asserting an explanatory relationship.[11]

To summarize: there is an area of common ground between Spencer and Dewey. Both saw living systems as systems whose interaction with the environment is characterized by maintenance of the organization of the living system. Both used the language of "equilibrium" in describing organism/ environment relations. Both thought the pattern of cognition has some formal similarity to the pattern of basic life-activities, more so for Spencer than for Dewey. And against this background, both understood the role of cognition in life in terms of adaptive response to environmental complexity.

4.5 Past and present

In the first chapter I distinguished two senses of function which have been isolated in philosophy of science, and I associated each with a construal of the environmental complexity thesis. First, the environmental complexity thesis can be read as a claim about what cognition is for, in the sense of its *raison d'etre*, the thing it does that explains why it is there. This is the teleonomic version of the thesis. But knowing what something is for is not the same as knowing everything it is *good for*. Another set of questions about the role of mind in nature comprises questions about instrumentalities, questions about how intelligence functions in the attaining of desired results by appropriate coordination of components.

Spencer's version of the thesis does not make use of a distinction of this type. For Spencer there is a single pattern visible in evolutionary history and individual development, in which cognition contributes in a particular way to the production and maintenance of relations of adjustment and correspondence between organic systems and their environments. Cognition's evolutionary role is adapting agents to environmental complexity, and this development is part of a general trend towards the greatest perfection and the most complete happiness. That is cognition's role in our lives, past and present. In Chapter 3 questions were also raised about whether cognition can acquire a function in the strict teleonomic sense from some of Spencer's "equilibration" processes, but that is not the issue here. The point is that Spencer did not distinguish between cognition's historical/evolutionary role and its instrumental role.

Dewey recognizes the difference between questions about origins and questions about present instrumental potentials. His view can be described roughly by saying that he accepts both a teleonomic and an instrumental version of the environmental complexity thesis, although this is oversimplifying in several respects.

In the first place, Dewey's views are complicated by his claim that nonlinguistic animals cannot think, as they cannot use symbols. As a consequence of this there can be no large-scale "evolutionary biology of mind" for Dewey, in the sense seen in Spencer and other writers. Thought is born in social interaction. The "antecedents" of thought, the biological structures and processes upon which thought and inquiry are built, do have an evolutionary history, and one that involves adaptive responses to environmental conditions. The brain is "the instrumentality of adaptive behavior" (1931, p. 214f).[12] But before evolution produces social interaction, this instrumentality of adaptive *behavior* is not an instrumentality of adaptive *thought*.

Once social interaction reaches a certain level of sophistication, symbolism and hence intelligence are possible. The advent of this form of interaction between agents may be the consequence of both biological and cultural evolution. However it is reached, from this point it is fair to attribute to Dewey a teleonomic version of the environmental complexity thesis. The hazardous quality of the world can be inferred from the existence of thought, because dealing with hazard is thought's *raison d'etre*.

Dewey's theory of inquiry is also supposed to have instrumental force. The problems posed by environmental uncertainty are not just part of the explanation for why cognition exists; they are also what makes thought a useful capacity, a capacity with a role to play in the explanation of agents' successful dealing with their environments. However, Dewey's position on the instrumental properties of mind also includes a role for thought that is independent of any links to behavior and environmental complexity. That is, the environmental complexity thesis is not the whole story about thought's instrumental properties for Dewey. The inner realm is not only an arena for planning and testing actions; it is also a "readily accessible and cheaply enjoyed esthetic field" (1929a, p. 227). Once minds exist, thought has its own recreational role and it can determine new types of interests, pleasures and goals.

Pragmatism is sometimes regarded as the view that seeking justified belief or any other epistemic property is only worthwhile because it helps in the attaining of behavioral goals. However well this might characterize the views of other "pragmatists," this is not a good description of Dewey's attitude. Thought and inquiry can be their own rewards, for Dewey, as can particular properties of thoughts or belief systems. It is true that Dewey thought that certain traditional epistemic goals are misguided. But the conception of epistemic value that should replace these superstitions has both a practical and an "esthetic" side (1929b, p. 111). If an agent values certain properties of thought for what they are in themselves, there is nothing wrong with that. For example, a certain type of aesthetic appreciation of thought – enjoyment of the problematic as problematic – is a quirk characteristic of the scientific mind (1929b, p. 182). To call this aesthetic is not to disapprove.

In general then, for Dewey, thought has three distinct roles. (i) Inquiry is a means to the solution of immediate problems. (ii) Gaining knowledge of nature is a laying down of tools and resources for the solution of unknown future problems. (iii) Thought is something that can be enjoyed for its own aesthetic properties.

Dewey's position on this question is a good one, as long as the term "aesthetic" is understood in a broad way. Pragmatism from some angles can look like the imposing of an austere work ethic on epistemic life – nothing is good in thought unless it is made to do some behavioral work. That is not in the spirit of Dewey. Thoughts with properties such as consistency and explanatory power, can also be sought for what they are in themselves. But if so, that is nothing more nor less than an aesthetic choice on the agent's part. It is not conforming to a higher epistemic mission or duty; it is not a choice which entitles the agent to a quasi-moral status that an individual who embraces inconsistent beliefs is not entitled to. The individual who chooses inconsistency may run into bad behavioral trouble, but that is another matter.

Thought has a legitimate aesthetic role in the agent's lives. Is this the *only* type of value that thought can have, when thought is viewed independently of behavior? On this point I am uncertain. There may be coherent views of epistemic value which are neither pragmatic nor aesthetic. On the whole I incline tentatively towards Dewey's view, in which the only respectable non-pragmatic value in thought is aesthetic. But this stronger view does not have to be defended here. The present point is that there is a real aesthetic role played by thought, of the type Dewey described.

In the sense outlined in Chapter 1, this aesthetic role is still an "instrumental" property of cognition – it is a contribution made to the well-being of some larger entity (the agent). So recognition of this aesthetic role of cognition is recognition of instrumental properties of thought that are not described by the environmental complexity thesis. Whether or not the teleonomic version of the environmental complexity thesis is central to the explanation of the existence of cognition, the environmental complexity thesis can only be *part* of the story about the instrumental role of cognition.

So this is a point at which Dewey's position is endorsed. The instrumental role of cognition need not always involve problem-solving, as properties of thought can be sought and enjoyed for themselves. But there is also an issue involving the distinction between teleonomic and instrumental where I dissent from Dewey's view.

As outlined earlier in this chapter, Dewey views inquiry as a response to indeterminate situations. This is a claim about the role of thought within problem-solving, as opposed to its aesthetic role, but, at least in Dewey's *Logic*, this is a claim about *all* problem-solving. Dewey wants to establish a link between some property of situations and the instrumental role of inquiry, even when inquiry is freed from the specific circumstances that can be held to have teleonomic relevance.

My view of this issue is based upon a sharp distinction between teleonomic and instrumental. If our project is teleonomic, there may well be a specific property of environments – complexity – that has a privileged role in the function of thought. But it is much harder to claim this on an instrumental version of the environmental complexity thesis.

Dewey's theory makes a condition of "indeterminacy" of situations into a necessary aspect of all genuine inquiry. Without indeterminacy there can still be thought, with its aesthetic properties, but there cannot be inquiry. This is too extreme a view. It is more reasonable to hold that though a great deal of inquiry may have to do with dealing with properties of this type, it is too strong to say this of *all* genuine inquiry.

The problem stems from the fact that on the instrumental version of the environmental complexity thesis, agents are not constrained with respect to their goals. They can determine their own aims and aspirations, and put thought to use in achieving results which may be far removed from teleonomically salient goals such as good health and successful reproduction. Instrumentally speaking, there is nothing stopping an agent putting thought to use in *avoiding* historically normal and teleonomically salient goals.

It is consequently over-strong to claim that all inquiry is a response to "indeterminateness" in a situation, unless this property is understood in a trivial way. If what is involved in indeterminateness is just a state in which the agent's relations to its situation are not the way the agent *wants* them, then this is a reasonable view but a weak one. The same is the case if indeterminacy is simply in the agent's ideas – if there is uncertainty in a subjective sense. These weak claims are different from the claim that inquiry is a response to some substantive environmental property of hazard, unpredictability or precariousness. The problem with the stronger claims is the fact that what the agent does not like about the situation could well be its very stability and determinateness. Boredom may not have an important teleonomic role, but it is not any less real for that. It is simply not the case, and once we reach sophisticated and independent agents there is no reason why it should be the case, that every intelligent agent "welcomes order with a response of harmonious feeling whenever it finds a congruous order about it" (1934, p. 20).

Consider an agent in a static, orderly and benign environment, where this very stasis is what the agent deplores and seeks to overcome. The agent's aim is to change dramatically; a change is desired for its own sake. It may be hard to do this in an environment that is completely stable. Considerable thought and planning, trial and error, may go into dealing with this situation. This is a case in which the situation is bad or unsatisfactory by the agent's lights; the situation poses a problem. The agent may be uncertain about what they can do to overcome their dismal stasis; there is uncertainty in the agent's mind. But the problem itself cannot be viewed as the consequence of hazard, precariousness, or indeterminateness, insofar as these are viewed as real properties of the situation rather than something imposed on it. It is not indeterminateness with respect to what environmental conditions "import and portend in their interaction with the organism" that is causing the problem (1938, p. 110). It is determinateness that is causing the problem.[13]

So I am skeptical about Dewey's attempt to describe a nontrivial property of situations which can be regarded as the spur to inquiry in all cases, including cases far removed from teleonomic importance. Here again, the environmental complexity thesis endorsed in this book is one that is less elaborate and ambitious than the versions in Spencer and Dewey. The instrumental version of the environmental complexity thesis is too strong, even within a problem-solving conception of thought. For once we leave the domain in which biological goals are at issue – once agents are free to pursue their independently determined interests – it need not be the complex or variable that is problematic.

There is another closely related aspect in which Dewey's claims about the relations between inquiry and environmental conditions are too strong. This issue will be discussed in the next chapter but I introduce it now because of its connection with the questions about indeterminateness. Dewey not only saw all inquiry as a response to the indeterminate, he also saw all inquiry as necessarily making an objective change to the situation, transforming indeterminateness to determinateness.[14] If it is true that not all inquiry is a response to indeterminateness, as claimed above, then it is also true that not all inquiry functions to change indeterminateness to determinateness. These might be typical features of inquiry, but they are not reasonably viewed as necessary features. We will return to this topic in Chapter 5.

4.6 Selection and the pattern of inquiry

Spencer, as discussed in the previous chapter, made extensive use of what are now sometimes known as "instructive" mechanisms of organic change. They involve the direct effect of the environment, as opposed to a trial-and-error "selective" pattern. Dewey objected to the passive view of living activity and intelligence seen in Spencer's view. Would it be fair to say that Dewey, influenced as he was by Darwin and James, had a selective rather than instructive conception of the role of the environment in thought? There is some truth in this idea, but the issue is complicated.

An important feature of Spencer's system is that little distinction is made between processes that operate within the life of the individual and processes that operate across generations. Spencer's "direct" mechanisms of equilibration of organism to environment are seen in both the day-to-day activity by which an individual learns about the world, and also the evolutionary development of the population. The "law of intelligence" describes both processes. As that law has a primarily instructive character, it is accurate to say that both Spencer's view of the cognitive life of the individual and his view of mental evolution were for the most part instructive rather than selective.

Dewey's view is different in several ways. First, he is not trying to engage in evolutionary system-building; he does not try to give a unified account of mental development within and across generations in the same terms. As far as the evolution of nervous systems goes, Dewey was (as far as I can tell) a straightforward Darwinian. In particular, he welcomed the open-ended and contingent nature of Darwinian (as opposed to Spencerian) evolution (Dewey, 1898; 1910b). As outlined earlier, Dewey is prevented from having a full-scale evolutionary biology of mind, charting the emergence of intelligent thought through the development of more and more complex animals, by his insistence that mind only exists in a social context. But insofar as he is a Darwinian about nervous systems, Dewey has a selective view of the evolutionary origin of the mechanisms underlying mind.

These evolutionary issues are not closely connected to Dewey's central epistemological projects. Dewey's primary epistemological aim was to describe the role of intelligent thought in the lives of human, social agents as they attempt to deal with problems. A theory of problem-solving and intelligence can be fairly described as a "selective" theory if it makes use of cycles of undirected variation and environmental selection of variants. Operant conditioning is the paradigm of a selective mechanism for within-generation psychological change.

Dewey's theory of problem-solving is certainly not a theory based on brute operant conditioning. As outlined earlier, in Dewey's view a crucial feature of intelligence is that it enables possible solutions to a problem to be tried out and assessed in thought without overt action. As Campbell (1974), Dennett (1975) and others have argued, it is still possible to usefully describe a process of this type as a "selection" process. But this is only so if certain conditions are met. Dewey's view of inquiry does not display the paradigmatic pattern of a selection process because the production of variants, the production of ideas that are possible solutions to the problem at hand, is not "undirected" or "blind" or "random." The production of possible solutions for Dewey is an *intelligent* process. An agent brings to bear on any problem a mass of past experience, a set of habits and expectations that determine which possible solutions are entertained.

Each conflicting habit and impulse takes its turn in projecting itself upon the screen of imagination. It unrolls a picture of its future history.... Choice is made as soon as some habit, or some combination of elements of habits and impulse, finds a way fully open. (1922, pp. 190–92)

There is some accuracy in viewing this as a "selection" process. Deliberation is an "experiment in finding out what the various lines of possible actions are really like" (1922, p. 190). A variety of options is assessed, and a choice is made according to the options' consequences. But the force of an explanation in terms of "variation and selection" is much reduced to the extent that the production of variants is intelligent and directed rather than random (Amundson, 1989).[15] The analogy stressed by James between spontaneously produced thoughts and biological mutations (section 3.7) is not a part of Dewey's theory of inquiry.

Although Dewey's view of intelligence does not have the paradigmatic structure of a process of variation and selection, there is one important point of similarity between Dewey's view and selection-based epistemologies. The role of the environment in Spencer's view of thought is as an *imparter of structure*. The pattern of experience determines the pattern of thought. The role of the environment in Dewey's theory, and also in some more strongly selection-based views such as Popper's, is as an *arbiter and tester*. The aim of thought for Dewey is not to take on the environment's pattern but to solve an environmental problem. The solution need not have structure in common with the environmental feature that posed the problem. The role of environmental structure is to pose tests and determine consequences; the environment "creates doubt, forces inquiry, exacts choice, and imposes liability for the choice which is made" (1929a, p. 421).

So although Dewey's epistemology is not wholly selection-based, the role of the environment in Dewey's theory is more akin to its role in a selective view than its role in some other views.

It would be easy to get the impression from this discussion of Dewey and James, and their differences from Spencer, that selective externalist views are always more moderate than instructive externalist views; that moving to a selective picture is moving towards internalism. This is a mistake. There is no necessary connection between the extremeness of an externalist position and its use of selective as opposed to instructive mechanisms. A selective picture can be highly externalist if it apportions explanatory weight between the "generate" and the "test" parts of the cycle in a certain way. If the production of variants is uncreative, omnidirectional, completely mechanical, and slight in magnitude at each step, while the environment as selective agent does all the creative, ordering work, then this is extreme externalism within a selective picture. Contemporary behavioral ecologists proceed on the assumption that a great variety of behaviors, good and bad, are produced in an undirected way, and the fitness-enhancing ones are preserved by selection (Krebs and Davies, 1987). There is no moderation of externalism here. Further, Spencer's Lamarckian view is strongly externalist, but other applications of Lamarckian evolutionary mechanisms are associated with *internalism*, as they involve heritable organic change made possible by acts of the organism's creative intelligence and will.

Dewey's externalism is more selective than instructive, and Dewey's externalism is also moderate: these are two independent points. Dewey's is a view in which the production of variants, possible solutions to a problem, is an intelligent and creative process. This is part of Dewey's more general stress on the active properties of mind; viewing mind as *responsive* to experience need not involve viewing it as *passive*. The environment is the source of problems and the arbiter of success, but it is not the only structuring factor in the explanation of inquiry.

4.7 Pragmatism and reliabilism

As it was originally expressed in Chapter 1, the environmental complexity thesis is a claim about the environmental conditions under which cognition is useful. If an environment is wholly regular and simple there is no need for the flexibility in action which cognition affords. One ambitious way of putting this is to say that the thesis asserts a lower bound on the complexity of environments in which cognition will evolve. If so, is there also an *upper* bound?

Whereas Spencer's account of the environmental properties that generate intelligence focuses solely on complexity, Dewey saw cognition's role in terms of a specific *combination* of environmental properties. A wholly simple world is a world in which cognition is useless, but in a wholly chaotic world useful cognition is impossible. Thus the environments in which cognition has a useful role to play are environments characterized by a combination of the fixed and the variable.

Dewey's own expressions of this idea tend to have more of a flourish:

> Unless nature had regular habits, persistent ways, so compacted that they time, measure and give rhythm and recurrence to transitive flux, meanings, recognizable characters, could not be. But also without an interplay of these

patient, slow-moving, not easily stirred systems of action with swift-moving, unstable, unsubstantial events, nature would be a routine unmarked by ideas. (1929a, p. 351)[16]

More straightforwardly:

The incomplete and uncertain gives point and application to ascertainment of regular relations and orders. (1929a, p. 160)

This idea in Dewey establishes interesting links between pragmatism and some contemporary theories of knowledge.

One of the leading approaches to analyzing knowledge in recent decades has been the "reliabilist" approach. This view holds that knowledge is distinguished from mere true belief by features of the physical, law-governed links between the belief and the state of affairs which the belief is about. Knowledge is distinguished by its being a true belief caused by a reliable process of belief-formation (Goldman, 1986), or by its being a true belief characterized by certain subjunctive relations between the fact that the belief is tokened and the fact that the belief is true (Armstrong, 1973; Nozick, 1981), or by its being a belief that *p* caused by the information that *p* (Dretske, 1981).

A central element in reliabilist epistemology is the idea that knowledge of external things is only possible if the world has a certain contingent character, if there really are highly reliable links between beliefs and the conditions in the world that the beliefs represent. If these links are never reliable, then knowledge is, contingently, impossible. (In some views, such as Dretske's, if there are no reliable links between inner and outer then *meaning*, as well as knowledge, is impossible.) If the world is sufficiently chaotic that states of affairs close to our peripheries, of the sort that affect perceptual mechanisms, give no information at all about what is going on further out in the world, then knowledge of those distal states is a false hope, and the best we can do is know what is going on at the periphery. Regularity in the world, of a certain type, is a prerequisite for knowledge.

This line of thought complements Dewey's. Dewey holds that environmental unpredictability is what makes cognition worth having. But he also holds that environmental predictability is what makes effective intelligence possible. It is true that Dewey differs from reliabilists such as Goldman and Armstrong in that they are concerned with common-sense standards of epistemic evaluation, while Dewey is primarily concerned with more pragmatic standards involving problem-solving and action. Nonetheless, this link between Dewey and the reliabilists has the potential to inform both pragmatism and reliabilism. Reliabilists discuss the correlations and regularities involved in knowledge in far more detail than Dewey does, while Dewey brings to the attention of reliabilism the need to explain not just what makes knowledge possible, but what makes it worth having.

If this point is accepted, many further questions arise. Does Dewey's view assert that there is a "window of opportunity" for intelligence and knowledge, between the point at which the world is so chaotic that knowledge is impossible and the point at which it is so dull that knowledge is useless? Or rather, is the reliability required different in *kind* from the unreliability required?

I am not concerned to work out exactly what Dewey meant – he was vague on this score. Instead I will describe one type of relation between environmental predictability and unpredictability which might be used to make sense of this idea, and develop the environmental complexity thesis in a promising way.

Cognition is useful in an environment which is characterized by:

(i) *variability* with respect to distal conditions that make a difference to the organism's well-being, and by
(ii) *stability* with respect to relations between these distal conditions and proximal and observable conditions.

This formula outlines a straightforward version of Dewey's claim that cognition is a means for using the "fixed" to deal with the "precarious." Cognition is most favored when there are (i) environmental conditions salient to the organism, which are not directly observable, and which are not stable or predictable in advance, and when there are also (ii) highly reliable correlations between these distal states and states which the organism can observe or detect more directly. "Observability" is understood here in a very broad way – many kinds of physical interaction with the organism will suffice. The organism can make use of these correlations to coordinate its responses to the variable states that affect its well-being. Without the unpredictable states in (i), cognition is not needed, and without the correlations in (ii), cognition cannot solve the problem.

These two conditions are not strictly sufficient for cognition being practically valuable. For example, some types of environmental complexity are such that there is nothing the organism can do to save itself, whether the environment's course can be predicted or not; some events imply a definite, rather than merely potential, "doom of disaster and death," to use Dewey's phrase. Within biology, some types of dramatically fluctuating environments are thought to favor very simple organisms: if it is very hard to survive bad seasons, the best option may be to have a capacity to reproduce very quickly when times are good (Bonner, 1988, p. 49).

These two conditions are not strictly necessary either. Condition (ii) is not necessary as even a completely unpredictable event can nonetheless be prepared for intelligently in some cases. Condition (i) is not strictly necessary as it may be a proximal condition, rather than a distal one, that is relevant and complex. It is probably unrealistic to hope for airtight necessary and sufficient conditions in this matter. But it is not unrealistic to think that (i) and (ii) do constitute an informative generalization about the value of cognition.

The problems posed by condition (i) are the raw material for pragmatist and evolutionary views of the mind, views that stress the problem-solving function of thought. Without real problems to solve, cognition's role can only be recreational. The regularities stressed by condition (ii) are the focus of some central products of modern naturalism – reliabilist epistemologies and indicator semantics. This is the core of the view of the mind proposed by people like Dretske. Without certain reliable links between proximal and distal, cognition has no purchase on the world.

To make this more intuitive, we can describe several types of environments in which cognition is *not* valuable.

First, the conditions in the environment that matter to the organism might be so stable that brute, inflexible modes of response are as good as anything else. There might be good cues in perception of what is going on, but nothing worth knowing about going on. The environment is then a pragmatic desert.

A second possibility is a combination of variety in the environment which is highly relevant to the organism, and an absence of reliable cues associated with these different distal states. There might be complex processes going on in the world whose outcomes causally affect the organism's well-being. But if the environment does not provide a way to track these conditions, cognition is useless. The environment would then, like a used-car yard, be a reliabilist swamp.

A third possibility – the worst possible world for cognition – combines both of these hostilities. The environment has little of interest going on, and the conditions that the organism can perceive are not even good guides to what is going on. Effectively similar states in the world generate diverse, unreliable perceptual states in the organism; the world is distally dull and proximally chaotic.

This third, worst case scenario is faced by the consumer of salad in many American restaurants. The available salad dressings are fairly similar, but the insignificant differences between them are marked with confusing and often unreliable names.

The reliabilist picture of the mind takes for granted the existence of things worth knowing, and describes how agents can make use of the resources offered by correlations and connections in the world to generate internal states that have special epistemic properties. It is central to pragmatism, on the other hand, to say that without practical problems that demand action, though the wheels of the mind might spin freely they are spinning idly. A real understanding of the place of mind in nature must incorporate both factors, must accommodate both lines of argument. This is a project which Dewey's *Experience and Nature* begins, and which the present book is intended to follow up.

If we describe an environment as one in which there is distal variability but a set of reliable correlations between proximal and distal, we are describing this environment from a certain point of view. What is distal to you might be proximal to me. But admitting this is not denying that the features described in (i) and (ii) above are real features of environments. Any environment has its own intrinsic structure; it has its own patterns of variation in temperature, its own media of transmission of light and sound, its own geography and so on. If, in virtue of this structure, the environment confronts the organism with distal variability in ecologically relevant properties and also a set of reliable correlations between proximal and distal, this is likely to be an environment in which effective cognition is both possible and worth having.

"Distality" in this sense need not involve large physical distance from the organism. One of the central types of environmental complexity faced by intelligent agents is complexity in the social environment. Some have argued specifically that this is the type of environmental complexity responsible for the evolution of high intelligence of the type found in primates and humans (Humphrey, 1976; Byrne and Whiten, 1988); this is the hypothesis of the "social function of intellect." The mannerisms, expressions and utterances of an agent with whom one is interacting are proximal; the behavioral dispositions of this agent in unusual situations are distal.

The ontological status of environmental patterns, and the role of the organism's point of view, will be discussed in detail in the next chapter. Before that I will describe some computer simulation work which provides an illustration of some themes from this chapter and the previous one, and I will then give a summary of the progress made so far in investigating the environmental complexity thesis.

4.8 A simulation

Todd and Miller (1991) explore the evolution of learning by modeling the evolution of the architecture of simple neural networks, using what is known as the "genetic algorithm." The neural networks can be thought of as marine animals which are born in the open sea, but which then settle down to an immobile life feeding on passing food particles. Once it has settled, the individual's problem is to decide whether to feed or not feed, when presented with each item of possible food. The environment contains both food and also inedible or poisonous particles, in equal proportions. When food is eaten the organism gains an energetic benefit, and when poison is eaten the organism pays a cost (though this is not fatal).

The particles of possible food have two sorts of properties that the organisms can perceive – color and smell. Food smells sweet and poison smells sour, but in this turbulent environment smells can mislead. The present olfactory input is only, say, 75 percent reliable as a guide to the particle that has to be eaten or rejected. That is, the probability of a sweet smell, given the presence of food, is 0.75 and the probability of a sour smell given poison is also 0.75.

The color of food is not affected by turbulence in this way, but color is unpredictable in a different respect. In half of the organism's environment food is red and poison is green, but in the other half the colors are reversed. Within each of these two micro-environments food color is 100 percent reliable.

The networks which Todd and Miller placed in this environment are constrained to have three "units," or nodes, only. These nodes are like idealized nerve cells. But there are lots of possibilities for their architectures. The units can be input devices of various kinds (red-detectors, green-detectors, sweet-detectors or sour-detectors). There is just one possible type of output unit (eating), and a "hidden" unit, which mediates between a detector and a motor unit, is also a possibility. The range of units an individual has is determined by its genetic make-up (genotype). The genotype also determines the nature of the connections between these three units. Connections between units are all constrained to be "feed-forward," or one-way. These connections can be hard-wired with either an excitatory or inhibitory connection, or they can be individually plastic. If the genotype specifies a plastic connection then the nature of the connection is determined by a simple associative ("Hebbian") learning rule. That is, if the two units tend to fire at the same time in the individual's experience, they acquire a positive connection between them – one unit comes to have an excitatory connection to the other. If they do not tend to fire together then the connection becomes negative or inhibitory. The question the model is intended to address is: when and in what ways will individuals with the ability to learn evolve in the population?

Each generation contains a large number of individuals of different types, which settle at random in the two different micro-environments and then make decisions about what to eat, based on their genotype and their experience. At the end of a fixed period they reproduce (sexually) according to their accumulated fitness, with the possibility of mutation and recombination of genes, and the new generation then settles in the environment at random and the cycle begins again.

At the start of the process the population consists of randomly configured individuals, most of which do not fare well. For example, some will not have a motor unit at all and will never eat, or will have a motor unit connected to an input unit which has the wrong setting – it might tell the organism to always eat when the present food particle smells sour. Another type of miswiring might be called "the academic." An individual can have two input units and a motor unit, but only learnable connections between all the units, connections which are initially set at zero. Suppose such a creature lands in a patch where food is red. Then it will learn the statistical association between redness and a sweet smell – the red-color input unit will tend to be on at the same time as the sweet-smell unit. But nothing is inducing the individual to eat. The motor unit will never be turned on, and its knowledge of the world will not do it any good, as far as nutrition is concerned.

Two kinds of wiring do make sense though. One useful wiring has a fixed positive connection between a sweet-smell sensor and a motor unit, and nothing else which influences behavior. This organism will generally eat when there is food present, its reliability being determined by the precise correlation between sweet smell and the presence of food. So in the present case, it will make the right decision 75 percent of the time. After a short period, these individuals tend to proliferate in the population.

The best possible wiring is a variant on this one, which has a fixed connection between a sweetness sensor and a motor unit (as above), but also a learnable connection between a color sensor and the motor unit. From the start this individual will tend to eat when there is food, as the smell sensor is controlling the motor unit. But in addition there will be a correlation between eating and some state of the color sensor. If the micro-environment is a red-food one, then when the organism eats it will also tend to be seeing red. The correlation establishes a connection between the color sensor and the motor unit, and (given the right initial settings) this connection will eventually be strong enough to control the motor unit by itself. Then the eating behavior will be controlled by a 100 percent reliable cue for the remainder of the individual's life. Typical runs of the simulation begin with the fairly rapid evolution of the simple, hard-wired smell-guided networks, and some time afterwards learners appear and take over.

This is a simple experiment but a philosophically rich one, and it illustrates several themes from the preceding discussion.

Chapter 1 distinguished between first-order and higher-order properties of plasticity. A first-order plastic individual is one which does different things in response to different situations. The hard-wired smell-guided networks which initially take over the population have first-order plasticity. They are plastic in contrast with an individual which has a motor unit always turned on (or off). In the environment used by Todd and Miller, food and poison come by with equal frequencies, so a permanently-eating architecture will have low fitness when compared to a first-order plastic individual that uses smell as a cue. Note that if almost everything in the environment was food, then the permanently-eating architecture would do well. If smell was an unreliable cue, then the completely inflexible architecture would do better than the first-order plastic one. This possibility will be discussed in more detail in Chapter 7.

The environment which Todd and Miller use is one that does favor first-order plasticity. It is an environment in which there is uncertainty about whether the present particle on offer is food or poison, and it is also an environment in which there is a reliable correlation between these distal states and states the organism can perceive. Smell is a good guide to something worth knowing about.

The same factors also explain the success of learning. In Chapter 1 I called learning a second-order plasticity. It is changing the conditionals that govern the organism's behavior in different circumstances. In the early stages of a learning individual's life, color has no effect on behavior, but during the course of its life the individual comes to eat in response to perception of a particular color. Learning wins out over mere first-order plasticity in this model as a consequence of the same type of environmental mix of the variable and the stable. For any individual there is uncertainty initially about whether food is red or green in its micro-environment. If almost every part of the environment was characterized by red food then it would not be worth taking the time to learn that food is red. Learning would not be favored. But again, Todd and Miller use a case where there is a large degree of uncertainty about this distal state: half the environment is one way and half is the other. Although the color of food is not predictable in advance, the past experience of an

123

individual is a good guide to the future. Within each micro-environment food is always the same color, and color is also a reliably perceivable signal. So the learner uses this stable aspect of the world to deal effectively with the uncertainty over the color of food.

It should be clear that in this situation the explanation for the success associated with plasticity has the form of the environmental complexity thesis. Patterns of complexity and stability in the environment explain the adaptive value of specific plastic strategies over more inflexible alternatives. One combination of variability and fixity in the environment favors the first-order plastic organisms over completely inflexible ones, and another combination of environmental variability and fixity favors second-order plasticity over first-order.

4.9 A summary of progress made so far

We have looked at two different versions of the environmental complexity thesis, and at the role played by these versions of the thesis in two general views of mind and nature. Both Spencer and Dewey embedded their claims about the link between mind and environmental complexity in detailed and ambitious theoretical structures. The version of the environmental complexity thesis I endorse is more pared down and less ambitious. At this point I will summarize the main features of a version of the thesis which emerges from the last four chapters as a promising one. This will be a preliminary summary, as we have yet to consider what I view as the most fundamental difference between Spencer and Dewey. That will be the topic of Chapter 5. But it will be useful to clear the decks first, and tie together the main ideas we have encountered so far.

There are teleonomic and instrumental versions of the environmental complexity thesis. The instrumental version of the thesis is certainly too strong as it stands. The environmental complexity thesis, no matter how broadly construed, can only describe part of the instrumental function of thought. Following Dewey, we should admit a nonbehavioral role for thought as well as a role in behavioral problem-solving. Dewey called this an "aesthetic" role for thought, and this is an appropriate term. To call this role "aesthetic" is not to denigrate it in any way.

Though Dewey admitted an aesthetic role for thought, he claimed in some work that insofar as thought is involved with problem-solving, we should view all intelligent inquiry as aimed at dealing with situations which are "indeterminate." Though it is hard to work out exactly what this means, I think there is no interpretation of the claim which is both nontrivial and true. Thought can play an instrumental role in the attaining of goals which agents value even though they are far removed from, and possibly contrary to, the standard, teleonomically important goals of individual survival, health, reproduction and so on. However "indeterminacy" is understood (short of understanding it as "undesirability") there is no reason why agents should only find indeterminacy problematic.

So there is a straightforward argument against a strong, literal statement of the instrumental version of the environmental complexity thesis. Consequently, this thesis must be weakened before it can be accepted: dealing with environmental complexity is one out of a range of instrumental properties of cognition.

There are no similar fast arguments against the teleonomic version of the environmental complexity thesis, which is an adaptationist hypothesis. But we can say that the teleonomic version of the thesis does not suffer from the particular problems we have discussed with the instrumental version. The teleonomic version of the environmental complexity thesis is not immediately affected by the fact that agents can pursue eccentric or historically unusual goals.

The teleonomic version of the thesis is an externalist claim, like many adaptationist claims. But it need not involve extreme externalism. In particular, it need not involve a conception of mind based on a passive accommodation of experience or environment.

On the whole, Dewey's version of the environmental complexity thesis is far superior to Spencer's. Spencer's view does have two advantages over Dewey's though. First, Spencer understood complexity primarily as heterogeneity. This is a simple concept which is relatively easy to apply in particular cases. Dewey made use of a range of terms to describe the conditions to which thought is a response, and it is much harder to work out exactly what these terms assert, and how to detect these properties in situations or environments. In developing the environmental complexity thesis it is best to stick with the simple conception of complexity as heterogeneity, as far as possible.

Second, I envisage the teleonomic version of the environmental complexity thesis as part of a large-scale theory of the evolutionary biology of mind. The use of perception, internal representation and behavior to deal with environmental complexity is not a capacity found solely in social and language-using creatures like ourselves. It should be possible to develop general theories describing how different animals do this, and how more complex forms of cognition emerge from simpler forms. Dewey is prevented from having a view of this type by his insistence that mind does not exist outside of social and linguistic interaction.

Spencer and Dewey both embedded their views of mind within general theories of the nature of living systems. Spencer asserted strong continuity between life and mind, and Dewey asserted weak continuity. Neither of these constitutive theses is required by the environmental complexity thesis, in either teleonomic or instrumental forms. Weak continuity is independently plausible. Dewey's particular reasons for denying strong continuity are not compelling, though strong continuity is still questionable. The environmental complexity thesis, especially in its teleonomic form, does require a commitment to the methodological continuity thesis.

Dewey's version of the environmental complexity thesis features a focus on selective mechanisms over instructive, with respect to the role of the environment. Dewey's conception of intelligence does not have the paradigm properties of a selection mechanism, but there is some similarity here. Spencer's position, on the other hand, is a paradigm of an instructive view.

There is a temptation to think that the rise of selective theories has shown that more direct, instructive externalist models are more primitive (Changeaux, 1985, pp. 279–81), and perhaps that explanations in terms of selective mechanisms are now preferable *a priori*. This is over-stepping. In some cases selective mechanisms have clear empirical superiority over instructive; Darwinism has better empirical credentials than Lamarck's view of evolution. Selective views of the within-generation adaptation of the immune system to diseases are better-supported than instructive views. But selective mechanisms are not always and necessarily better than instructive; this is a decision to make case by case.

The teleonomic version of the environmental complexity thesis does itself require the existence of a selection mechanism, in a broad sense. It is through selection of various kinds that teleonomic functions are assigned, so selection must figure in the explanation of cognition's existence or maintenance, for cognition as a whole to have a teleonomic function. But this does not imply that the day-to-day operation of cognitive mechanisms must be strongly "selective," in the sense seen, for example, in mechanisms of operant conditioning and Popperian philosophy of science. There is no need to replace Spencer's global theory of mental development with a new, global, selectionist theory of mental development at all levels and time-scales. The selection mechanism of Darwinian evolution, the mechanism that originally gave rise to cognition, need not produce selection devices. It can do so, but it need not. Selection might be more important in the phylogeny of cognition than it is in its

126

ontogeny. The version of the environmental complexity thesis developed in this book does not require any particular account of *how* cognition works. No specific theory of learning and problem-solving, selective or otherwise, is required. For such theories we look to psychology and cognitive science. The environmental complexity thesis claims, in its teleonomic form, that a basic mental tool-kit has the function of enabling agents to deal with environmental complexity: that is why the took-kit is there. This claim is compatible with a range of theories of exactly how thought manages to perform this task.

Lastly, one of the best ideas in Dewey's version of the environmental complexity thesis is his claim that thought has its value in environments which exhibit a balance between the "precarious and the stable," or, more simply, the fixed and the variable. Once this idea is sharpened up, it has great promise as a unifying idea in epistemology.

Notes

1 *Experience and Nature* is thought by some to be the most "idealist" of Dewey's major works, the work which betrays his continental influences most clearly. This may well be so, but *Experience and Nature* is also a work which is more readily linked to contemporary naturalistic philosophy than some of his other works. *Experience and Nature* is less burdened with empiricist strictures than *The Quest for Certainty* (1929b), and it is not as bound up with unusual views about propositions and so on as *Logic* (1938). Rorty (1977) singles out *Experience and Nature* as something of a metaphysical lapse on Dewey's part; in my view it is indeed metaphysical, but very far from a lapse.

Burke (1994) gives a naturalistic reading of Dewey which is focused specifically on Dewey's *Logic*. This work is Dewey's most detailed epistemological statement, so it will be discussed in this chapter. However, many of the points at which I dissent from Dewey concern issues stressed in *Logic*. For Burke, the key to making Dewey's basic ideas plausible is stressing the integration and inseparability of organism and environment – this makes many of the initially strange ideas in *Logic* make sense. In my own view, some of Dewey's most valuable ideas are obscured unless we do distinguish between changes made to the organism and changes made to the environment, and it may not be possible to salvage everything Dewey said in *Logic*. Burke might regard this as residual "dualism" on my part.

Of course, many aspects of Dewey's views evolved over his long carrer. Whenever possible, I focus on his work in the 1920s, as I think it is the best work he did on these topics. So in some cases, when I say that "Dewey" made a particular claim, I strictly mean "Dewey of the 1920s." The references should help to make this clear.

There are many difficult issues of interpretation surrounding these parts of Dewey's philosophy. I will raise some of the problems in later notes to this chapter.

The notes to this chapter are mainly intended for readers with an independent interest in Dewey.

2 Dewey's use of contingent properties of environments in his theory of inquiry is also a feature which distinguishes him from his fellow pragmatists, Peirce and James. Both Peirce and James did think the universe had a general character of chanciness or uncertainty. At least, Peirce asserted this explicitly while James inclined to believe it and defended the right to believe it. Peirce stressed chanciness while James stressed a property of "unfinishedness" which I take to be related. But neither thinker began their *epistemology* in a link between these specific properties of the world and the function of thought.

3 Here, as I said earlier, Dewey is prone to using the concept of equilibrium loosely. In the *Experience and Nature* discussion, need is not a state of disequilibrium or disturbed equilibrium (1938) but instead a state of "uneasy or unstable equilibrium" (1929a, p. 253).

4 There is a difference between the view of life Dewey presents in *Experience and Nature* (1929a) and that in *Logic* (1938). In his 1938 presentation Dewey says that it is not strictly accurate to say that life-activities are defined by maintenance of some characteristic equilibrium *state* of the organic system, but rather by maintenance of a characteristic *relationship* between system and environment (1938, pp. 34–35). One aim Dewey has here is to enable development by the organism over its lifetime to be regarded as part of the same basic self-maintaining properties of life as the more obvious "conservative" structure-preserving activities. Development can be regarded as part of the maintenance of a set of relations to environment.

I am not sure if this is better or worse than the simpler view, where the living system itself is what is said to be preserved, by means of interaction with environment. See Burke (1994) for a reading of Dewey which places great emphasis on integration of organism and environment of the type asserted in *Logic*; inquiry is not undertaken by an organism, but by the organism/environment system as a whole.

5 Dewey used the term "continuity" continually, and he seems to have meant a number of different things by it. Phillips (1971) and Burke (1994) distinguish a "vertical" sense from a "horizontal" sense of continuity. The former is genetic or historical – it has to do with the relations between inquiry and other organic activities in the past. The latter has to do with the relations between different things existing at one time (or considered independently of time, perhaps). Both issues are important for Dewey.

Strong and weak continuity between life and mind, of the type outlined in Chapter 3, are both "horizontal." Both are strictly speaking independent of issues about evolution and the past, although both are naturally associated with certain historical claims.

See also Goudge (1973) on mind and continuity in Dewey.

6 This is a central feature of Dewey's view of intelligence (1922, p. 190; 1929a, p. 291; 1929b, p. 18; 1938, pp. 63, 111). For a more detailed discussion of the difficult topic

of Dewey's view of the relation between mind and the social, see Tiles (1988) Chapters 3 and 4.

7 See Stalnaker (1984) for both defence of a naturalistic theory of this type, and for a discussion of the issue of the content of nonhuman animals' thoughts. I am in agreement with Stalnaker's views on animal belief, and on the deficiencies of what he calls the "linguistic picture" of representation. I am not as optimistic about the specific theory of meaning he proposes, which is an atemporal indicator theory. Naturalistic semantic theories will be discussed in more detail in Chapter 6. See also Stich and Warfield (1994) for a good survey of the range of theories available.

8 There are many issues that would need to be addressed here. "Uncertainty" can be understood in objective ways and in more subjective ways. It is important also that there are two senses of "changeable," or "unstable," or "precarious" here. Some things in the world change a lot, of their own accord. Others can be changed by *us*. There are naturally static conditions that are nonetheless easily altered by us (ozone layer) and there are naturally variable conditions that we can do little to alter (night follows day).

9 When a situation is regarded as doubtful or indeterminate though it is really not so, this is even regarded by Dewey as "pathological" (1938, p. 110).

10 Here is the official definition of inquiry in Dewey's *Logic*.

> Inquiry is the controlled or directed transformation of an indeterminate situation into one that is so determinate in its constituent distinctions and relations as to convert the elements of the original situation into a unified whole. (1938, p. 108, italics removed throughout)

A comparison to Spencer is illustrative. Spencer's law of evolution (section 3.1 above) involves a change from the "indefinite" to the "definite." Dewey's account of inquiry involves a transition from indeterminate to determinate. The properties are similar, but the mechanisms are very different. Spencer's change will happen regardless of what individuals choose; the universe will bring it about. Dewey is here describing a paradigm of deliberate and *optional* human activity.

11 An important passage for this exegetical question is *Logic* (1938) p. 110: biological conditions of imbalance between organism and environment are the "antecedent conditions" of genuine unsettled situations. Environmental variability does produce these conditions of biological imbalance, so the issue then is how we should interpret this relation of being an "antecedent condition."

12 See also Dewey (1917) p. 78; (1929a) p. 23.

13 This topic is an especially complicated one. See Burke (1994) for an account of Dewey's view of indeterminate situations which attempts to both stress the biological underpinnings of the concept (as I have) but also to justify the idea that the concept can be used very generally in characterizing inquiry. Burke, for example, accepts that even the problem faced by a Ptolemaic astronomer in

predicting planetary motions involves "some sort of local disequilibrium in organism/environment relations" (1994, p. 147).

See also Thayer (1952) for a more guarded treatment of indeterminate situations in Dewey's theory.

14 Again, this is what Dewey claimed in his *Logic* (1938), at least. (See note 10 above.) This is another of the claims stressed specifically in *Logic* that I have problems with.

15 I would say that Amundson's general view is correct here:

> As we proceed "higher" on the hierarchical stages from evolution to psychology to social or scientific development, we should expect the explanatory force of naturalistic selection to gradually disappear. This is an ironic result of selection's own success at lower levels. The variation has been reduced, and in some cases directed, by selective processes at each lower level. (1989, p. 430)

Dewey (1898) makes an analogy between selection across generations and selection within the life of the individual by the cultural forces of public opinion and education. There is again no stress on "random" variation here, however. He also notes that natural selection will select for efficient modes of variation.

16 There are some problems of interpretation in this area. I take Dewey to be saying that without irregularity, there would be no point in thought's existence, it would be an idle wheel if it did exist. This idea is also found in a discussion of science:

> Science seizes upon whatever is so uniform as to make the changes of nature rhythmic, and hence predictable. But the contingencies of nature make discovery of these uniformities with a view to prediction needed and possible. Without the uniformities, science would be impossible. But if they alone existed, thought and knowledge would be impossible and meaningless. The incomplete and uncertain gives point and application to ascertainment of regular relations and orders. (1929a, p. 160)

Sometimes Dewey appears to mean something stronger than the claim I attribute to him in the text. He may mean that without irregularity it is impossible for thoughts to even occur. "Unless there was something problematic, undecided, still going-on and as yet unfinished and indeterminate, in nature, there could be no such events as perceptions" (1929a, p. 349). If thoughts and perceptions are *defined* in terms of their role in dealing with problems, the two readings collapse into one. But this does not fit with the passage discussed earlier in which thought is accepted as an "esthetic field." I leave the exegetical point to more experienced Dewey scholars.

5

On Construction

5.1 Asymmetric externalism

In this chapter we will look at a critically important issue, which marks what I view as the most important divide between Spencer's version of the environmental complexity thesis and Dewey's version. This issue is the role played by organic systems, such as intelligent agents, in the construction of their environments.

Up to this point we have been primarily concerned with different approaches to explaining organic properties, and with the idea that cognition, a form of organic complexity, might be understood in terms of its relations to environmental conditions. Our perspective has been "outside-in." In this chapter we will look down this same road from the other direction; we will look at explanations of environmental properties in terms of organic properties. The direction of influence we are concerned with now is "inside-out."

We will approach these questions, as we did in Chapter 2, by looking first at the relationships between some basic explanatory schemas, and describing a range of disputes in different fields in terms of this categorization.

In Chapter 2 I distinguished three basic types of explanation: externalist, internalist and constructive. Externalist explanations explain internal properties of organic systems in terms of environmental properties. Internalist explanations are explanations of one set of internal properties in terms of another, and *constructive* explanations explain environmental properties in terms of organic properties. Some externalist explanations are c-externalist: they are externalist explanations where explanandum and explanans are both properties of complexity. These categories, again, are idealized. Adaptationism is generally externalist but adaptationists do not deny that internal factors such as genetic and developmental constraints exist and make some difference. It is a question of the relative weight associated with different explanatory factors.

Externalist explanatory programs in the sense of Chapter 2 assert one direction of explanatory relevance – from environment to organism – but they do not necessarily deny the importance of the other direction. That is, I understood externalists as giving positive theories about how various organic properties depend on environment; externalists were *not* viewed as needing to take any particular stand on whether the state of the environment depends on the activities or properties of the organic system. So giving an externalist explanation of organic property O_1 in terms of environmental property E_1 does not preclude giving a constructive explanation of E_1. E_1 might be the product of some other organic property O_2.

So let us now define a stronger "externalist" view, which will be called *asymmetric externalism*. An explanatory program is asymmetrically externalist if it (i) explains properties of an organic system in terms of properties of the system's environment *and* (ii) explicitly or implicitly denies that these properties of the environment are to be explained in terms of other properties of the organic system.

A fundamental difference between Spencer's and Dewey's versions of the environmental complexity thesis is that Spencer's view of mind is largely an asymmetrical externalist view and Dewey's is not. Dewey's view of the place of mind in nature includes, as an essential part of the picture, a role played by mind in transforming the world. It is a two-way view. A version of Dewey's claim will be accepted in this book, as part of the most promising way to develop the environmental complexity thesis, in both teleonomic and instrumental forms.

Before looking at Dewey's position we will consider the general features of asymmetric externalist views in more detail.

An asymmetric externalist view of an organic system asserts that while the organic system is controlled, in some respect, by environmental properties, the environment is either fixed or is governed by its own intrinsic dynamic. The environment goes its own way and the organic system follows. A point I will constantly return to in this chapter is that there are several distinct ways in which this view can fail to be true in a particular case.

(i) First, the properties of the organic system might be explained in terms of other properties internal to the system, where the same is true of the environment. Each evolves according to its own dynamical principles;

each follows the beat of its own intrinsic drummer. This can be called a *decoupled* view.

(ii) Alternately, we could explain the properties of the organic system in terms of environmental properties, but then turn around and explain those properties of the environment in terms of other properties of the organic system. This need not involve a "circularity," in the pejorative sense. Both externalist and constructive patterns of explanation apply, with different phenomena being explained. This type of case will be discussed in detail in this chapter. Views of organic systems that combine these two patterns of explanation will be called *interactionist* views.[1]

(iii) Lastly, it might be that the organism develops according to its own internal principles and also determines what happens to the environment. Then internalist and constructive explanations both apply. This view, which also asserts an asymmetry, can be called an *asymmetric constructivist* view.

An example may help. Consider a crowd at a football game. The level of noise made by the crowd at some time can often be explained in terms of something happening on the field. A loud burst of sound might be the consequence of, and explained by, the home team scoring. A precarious, unpredictable game will usually generate more noise than a walk-over. So there is explanatory salience in that direction, from the game to the crowd. It is an empirical matter of how closely the crowd's behavior is tied to the activity on the field; either they are watching and reacting or they are not.

If we take the spectators as analogous to the organic system and the game as analogous to the environment, then externalism here is true, or the externalist picture applies, if the crowd really is tracking the events on the field closely and responding to the game. At a cricket Test match (which lasts five days) there is often a looser link between what the crowd does and what the players are doing. Much cricket crowd noise is generated endogenously and has its own intrinsic dynamic properties. So this is a case in which a more internalist view of the crowd's behavior is appropriate.

But suppose we are not concerned with a cricket Test match and the game does influence the crowd. Then there is the further question of the influence of the crowd on the game. To what extent does the noise emitted by the crowd affect the performance of the players? Are they inspired by their supporters or are they unable to even hear what is happening in the stands? There is an empirical question about how much the effect of the crowd contributes to a home-team advantage in each particular case.

Continuing the analogy, an asymmetric externalist view is true in this case if the game has a great effect on the behavior of the crowd but the crowd does not make much difference to the players. If the crowd and players generate their own patterns of behavior internally, or if the crowd and the players both affect each other greatly, or if the players are slaves to the intrinsically determined behavior of the crowd, then asymmetric externalism in this case is false.[2]

Let me stress that these are *distinct* ways in which an asymmetric externalist program can fail. It can fail as a consequence of the role played by the intrinsic structure of the system in filtering or transforming the inputs it receives. To the extent that this occurs, knowing how the world impacts upon the system is not knowing what you need to know to predict what the system will do. Environmental inputs may function as a mere fuel or trigger, rather than a directing agency. The debates discussed in Chapter 2 were debates about this issue – the very viability of externalist explanatory programs. The behaviorist quest for a theory of learning formulated entirely in terms of reinforcement is an attempt to say what will happen without attending to the particular internal structure of the system. Arguments in psycholinguistics from the "poverty of the stimulus" assert that this program fails, at least in some domains. In biology, adaptationists attempt to devise theories formulated in terms of phenotypes and environmental pressures, which generalize over a variety of underlying genetic mechanisms. In reply, geneticists and developmental biologists have attempted to show the important role played by various internal details of evolving systems. These are arguments against externalism, whether the externalism is asymmetric or not.

Alternately, the externalist part of the explanation may work, but there may be a two-way coupling between organic system and environment. Environmental inputs in this case do largely determine what the organic system does, but the actions of the organic system then turn around and impact upon the environment. The externalist part of the explanation is intact, but there is no asymmetry in the roles played by each party.

An example of this possibility is furnished by some events early in the evolution of life, as Lewontin has stressed. The oxygen-rich atmosphere around the earth was not a fixed background against which life evolved, or one whose development was independent of life. Instead it was the product of early stages in organic evolution. Once this oxygen-rich atmosphere is in place, the course of subsequent evolutionary processes is greatly affected. But the continuing nature of this atmosphere is still dependent on the actions of living beings; the critically important role the atmosphere plays in determining what living beings are like does not negate the critically important role living beings play, in determining what the atmosphere is like.

It is worth stressing again the narrow way in which the term "constructive" is used in this book. Many writers use the language of construction when all that is intended is an internalist point: "theories are constructed by scientists, not imposed on them by data." This is not a constructivist claim in the sense of this book unless it is also claimed that the world, or the environment, is constructed. As long as the object of construction is part of the organic system itself (phenotype, theory, language, judgement...) the explanation is internalist.

A fundamental point of this chapter can be expressed here as a dictum: *Distinguish between internalist arguments and constructive arguments against asymmetric externalist positions.* Do not confuse the self-absorbed crowd at a cricket match with the football crowd whose oppressive chanting causes a player to miss a kick at goal.

5.2   Two lines of dissent

In Chapter 2 I listed a range of theoretical programs in different fields and classified them as externalist. These included adaptationism, empiricism, associationism, and so on. Is it also true that these are asymmetric externalist programs? That is in many cases a difficult question. It is difficult firstly because different empiricists, for example, could differ on the question of asymmetry even when they agree about the dependence of thought on experience. But more importantly, it is difficult because in my definition of asymmetric externalism I said these views include an explicit or *implicit* denial of a dependence of the environment on the organic system. It is fair to say that many externalist theories do not say much about the dependence of environment on organism, but when does this constitute an "implicit denial"? It is not possible to explain everything at once, and the externalist can fairly say that the fact that associationism, for example, does not make a big issue of the effect of agent on environment does not amount to claiming there is no effect.

The case of classical empiricism is complicated here. Locke and Hume are hard to classify, as they do not so much deny that agents affect their environment but do not much discuss it. They are concerned to give a positive theory of what happens to agents as a consequence of experience. Berkeley is different, as he had an explicit theory of what is going on outside the mind determining the ongoing pattern of input. The course of experience is directed by God, and it is to God's benevolence that we owe the "steadiness, order, and coherence" of our sensory input (Berkeley, 1710, sec. 30). This is an "environment" for the mind which is not very open to human transformation. God is self-sufficient if anything is. (At least, He is self-sufficient in this respect except insofar as He is bound by policies He takes towards the decisions of human agents.) So Berkeley's view might be regarded as asymmetrically externalist.

It is hard to know how to treat views that do not deny the effect of the agent on the world but systematically neglect it. Empiricist views have detailed categorizations for the different types of effects experience can have on the mind. They do not take these pains with the different types of effects the mind can have on its environment. Some empiricists have such epistemological scruples that many investigations of this kind would be regarded as impossible, in fact. However, if we look at the *overall* world view of many empiricists, such as J. S. Mill, it could hardly be said that they were hostile to projects involving making changes to the world and society.[3]

In any case, I will not make more specific judgements about exactly where the "implicit denials" characteristic of asymmetric externalism might be found in empiricism and related psychological programs. This would require more detailed historical work.

The case of adaptationism in biology is also difficult. For the most part adaptationists can again only be accused of neglecting rather than denying the influence of organic systems on their environments. I understand one line of argument in Lewontin's critique of adaptationism as the claim that a sufficiently systematic neglect of organic action on the environment is as good as a denial of it (1983). Further, the paradigms of adaptationist explanation are often cases where the asymmetry is sharp. There is no more celebrated case of selective explanation than the effect that air pollution in England had on the coloration of moths which are preyed upon by birds (Kettlewell, 1973). The trees became darker, and the moths followed. It might be argued that when a case like this is taken as a paradigm, a certain picture of the relative importance of the two possible directions of influence is suggested, and this view becomes embodied in everyday scientific practice.[4]

Although there are good examples of adaptationist work which suggest an asymmetric picture of organism and environment, there are also cases which suggest the opposite view. One of the most active areas within the adaptationist approach to biology is evolutionary game theory (Maynard Smith, 1982). If this program is regarded as externalist, it cannot be regarded as asymmetrically externalist. The "environment" relevant to an explanation of behavior in terms of evolutionary game theory includes the current distribution of strategies within the population itself. The behavior of any individual contributes to the properties of the environment of other individuals. This affects how those other individuals should and will act, and hence it affects what behaviors the original individual will encounter in the future. Of course, if the population is large and individuals play against each other only once, any individual will have a very small effect on what it is best for the others to do. Game theory can concede that you often can't swim against the tide, or alter its direction. That is not the point. The point is that the theory explicitly considers both directions of influence. The same is true of work on biological "arms races," and other co-evolutionary models investigated within the adaptationist framework (Dawkins, 1986, Chapters 7 and 8).

Some problems that arise here are related to problems discussed in Chapter 2, concerning nonstandard divisions of organism and environment. When beavers build a dam this is naturally regarded as constructive action on their environment ... unless the dam is considered part of the beaver. If we view a dam as an "extended phenotype," produced by the beaver's genes, for example (Dawkins, 1982), then this change might be viewed as one made to the organic system rather than to its environment. In Chapter 2 I said that Maturana used an extension of the organic system outwards to avoid an externalist concession. This same unorthodox boundary can also have the effect of reducing the role of *constructive* relations.

So the question of the extent to which externalist programs such as adaptationism and empiricism are asymmetrically externalist is often difficult. The more important point here is that asymmetric externalism *can* be discarded without discarding externalism generally. An adaptationist might accept with Lewontin that there is in many cases a symmetry in the roles played by organism and environment, while, if desired, holding to the view that natural selection is ever-present and largely unconstrained by internal factors. If adaptationism is an asymmetrically externalist program, there are two distinct ways in which it can give ground.

The same is true of empiricism, to the extent that empiricism is asymmetrically externalist. There are two distinct avenues of retreat or dissent. One can deny that thought is so closely dependent on inputs from outside, claiming that the mind has its own rules and patterns and goes its own way. This is internalism, and when applied to empiricism it generates a *rationalist* line of dissent, and also other internalist views. The other option is to accept that thought has very close links to experiential input, but then to assert a two-way connection between thought and the world. Thought is a response to experience, but in turn it is an instrument which transforms the world and hence the future course of experience. This is a *pragmatist* line of dissent. Pragmatism is an interactionist view of the mind.

So in this picture a pragmatist is analogous to the adaptationist biologist who yields to Lewontin on the role of the organism in transforming the world, but does not yield to people like Goodwin on the importance of internal developmental principles in determining the course of evolution.

When I say that this is a pragmatist idea, I have Dewey in mind more than anyone else. James did make claims of this type, but did not develop the idea in the detail seen in Dewey. James, especially in his early work, also showed as much sympathy to internalist lines of dissent (although not "rationalist" ones). This is illustrated by James' criticisms of Spencer, discussed in Chapter 3. I find Peirce hard to classify. It was Dewey who developed a detailed epistemology based upon a coupled or interactionist conception of the relation between mind and nature.

If this categorization of Dewey and (earlier) James is right, then it could even be said that within critiques of orthodox empiricism, James and Dewey play similar roles to those played by Gould and Lewontin in their attack on adaptationism. Gould and James, on one side, tend to oppose the asymmetric externalist by opposing externalism in general. James stresses "interests" internal to the mind which filter experience, and Gould stresses the role played by developmental and architectural properties in evolution. Dewey and Lewontin, on the other hand, focus their criticism on perceived asymmetries in empiricism and adaptationism. Dewey and Lewontin both view "construction" as a basic activity of organic systems.

In saying this I am treating Lewontin's focus on two-way interaction between organism and environment as the primary thrust of his critique of adaptationism. This is not to deny that Lewontin endorses some internalist arguments against adaptationism (1974a). In fact, I will argue later that some of Lewontin's officially constructivist arguments are better viewed as internalist. However, I think it is fair despite this to regard Lewontin's focus on construction as central, and hence fair to see a real analogy between his work and Dewey's. So orthodox empiricism was confronted with a two-pronged attack around the turn of the century which is similar to the pitchfork that adaptationism has had to deal with more recently.

We should now return to the environmental complexity thesis. Spencer's version of the thesis was, on the whole, part of an asymmetrically externalist conception of life, mind and humanity. I say "on the whole" because there are specific parts of Spencer's view that are not asymmetrically externalist, as discussed in Chapter 3. But admitting that Spencer was not at the most extreme point possible, with respect to asymmetric externalism, is not to deny that there is a vast difference between Dewey and Spencer on this score. Dewey was opposed to the idea that in solving problems organic systems merely accommodate environmental demands. Rather, they intervene in the environmental processes which generated the problem, and alter the environment's intrinsic course: "Adjustment to the environment means not passive acceptance of the latter, but acting so that the environing changes take a certain turn" (1917, p. 62; see also 1948, p. 84). This is the core difference between Spencer and Dewey's theory of living systems, and one that ramifies through their views about mind and knowledge as well.[5]

The environmental complexity thesis is, when understood in a teleonomic way, an externalist position. But this classification leaves open the question of whether it is asymmetrically externalist. Spencer's version was for the most part asymmetric. Dewey's was not asymmetric at all. As I have been stressing, it is possible to deny an asymmetrically externalist view either by adopting an internalist explanation for the organic properties in question, or by accepting a constructive account of the relevant properties of the environment. In the case of cognition, to take a strongly internalist view of the function of cognition is to give up the environmental complexity thesis. To accept the existence of constructive relationships is not.

I realize that using the term "externalist" at all about a view which admits two-way connections between organism and environment has the potential to mislead. But it is important to have a category which captures what there is in *common* between the various versions of the environmental complexity thesis we have looked at, just as it is important to appreciate the differences.

The division between Spencer and Dewey on the issue of asymmetry is a fundamental one because some of their more detailed disagreements can be understood as consequences or expressions of this more basic difference. In the next chapter the question of whether correspondence should be viewed as a real and basic relation between internal and external will be addressed in this light. Spencer was an enthusiast for correspondence – one of the most ardent enthusiasts that concept has had. Dewey thought that correspondence as an epistemic ideal promotes a passive, spectator-like attitude towards knowledge.

Similarly, Spencer thought that progress is inevitable. He saw it as a consequence of the most basic laws governing the universe. The way he generated a guarantee of progress in the organic world was by shackling organic systems to their environments, and holding that environments contain the right dynamic properties to produce directional change of the required type. Organic systems, including human societies, do not have to pull this train along, and do not even have a choice whether or not to ride it. Progress will simply occur and the only positive thing we can do to help is to refrain from interfering. Dewey's view on this issue was very different. Dewey saw progress in the world as something intelligent agents have to actively bring about, through fallible and practical attempts to deal with their problems. We cannot hope that the universe as a whole will automatically take us in a desirable direction. Improvement has to be achieved, and the only progress we can expect to see will be the consequence of an active transformation and reconstruction of our conditions.

The version of the environmental complexity thesis defended in this book sides with Dewey on the basic point about asymmetry. Cognition enables agents to deal with complex environments, but an important aspect of this process is transformation of environmental conditions.

There are some points of disagreement between certain versions of Dewey's position and the view defended here, however. Dewey in his *Logic* (1938) does not just say that transformation of the world is an important function of thought; he insists that *all* inquiry involves making changes to situations. All inquiry affects a "transformation and reconstruction of the material with which it deals" (1938, p. 161). It is difficult to interpret this without trivializing it or making it unreasonably strong. Thayer, for example, asks how a child who successfully makes his way out of a maze has transformed the material of the problem he faced (1952, pp. 176–184).[6] If the child burned the maze down, that would be a definite transformation. But if he walks through it leaving the walls intact, the problem has been solved but how has the "material" which generated the problem been transformed? The child has been transformed, but not the maze. If the child is part of the "situation" (a view which Dewey does hold) then perhaps the change to the situation is just the change to the child. But then the claim about inquiry's role in changing the "material" of a problematic situation is a very weak one.

I agree with Thayer. There is a great deal to be gained by recognizing in epistemology that intelligence acts in the transformation of the world. But only some problem-solving has this character, while some does not. Sometimes the only change made is internal to the organism or agent. It is not a return to a "subjectivist" or "dualist" view to hold that transformation of the situation or environment is a common, rather than necessary, feature of the problem-solving action of intelligence.

We should accept Dewey's claim that thought functions in changing the world, and that this constructive activity is as philosophically important as the *re*active properties of thought. But we should not accept a version of this idea which is too strong, such as the version in Dewey's *Logic*. We should accept a narrower construal of the constructive activity of intelligence.

5.3 Biological constructivism

The remainder of this chapter will discuss this narrow view of the constructive activity of intelligence, and of organic action in general.

There is a variety of relationships between organisms and their environments which might be regarded as "constructive." The version of the environmental complexity thesis defended here is a version which accepts one type of claim about the mutual dependence of organism and environment but rejects others. A number of biologists and philosophers have thought that the right view of organism-environment relations is a more thoroughgoingly "constructivist" view. The strongest forms of this position have similarities to the broader metaphysical antirealist position called "constructivism," the view that thought, or language, or scientific activity, constructs the external world. I will argue against stronger versions of constructivism in biology, by means of close attention to the difference between internalist arguments and constructive arguments against asymmetric externalist positions. A narrow conception of "construction of environments" will be outlined, and the constructive activity of thought in the world will be viewed as a particular case of the more general phenomenon of organic construction of environments. Then I will export some morals of this discussion and apply them to metaphysical disputes about realism. In particular, I will claim that arguments against asymmetric externalism are often mistaken for arguments against realism.

We will begin by looking at a set of arguments intended to motivate a strongly constructivist view of the relations between organisms and their environments. These arguments are found in the work of Richard Lewontin (1982, 1983, 1991; see also Levins, 1979).

For Lewontin, the imposition of an asymmetric externalist explanatory regime on biology was one of Darwin's central achievements. This enabled evolutionary theory to advance beyond (internalist) views in which changes undergone by species were understood as analogous to the unfolding of a developmental program. Darwin located the ordering mechanism of evolutionary change in the environment instead. One aspect of Darwin's imposition of an externalist explanatory framework on biology is what Lewontin refers to as the "alienation" of the organism from its environment; the organism is the passive object upon which external forces operate.

According to Lewontin, Darwin's asymmetric externalism was enormously progressive, when compared to earlier views. But it is not a true doctrine, and evolutionary theory has passed the stage when it is a fruitful assumption to make. Lewontin proposes that the standard conception of organism–environment relations be replaced by a picture in which organism and environment are each dependent on the other, both causally and ontologically.

> [T]he organism and environment are not actually separately determined. The environment is not a structure imposed on living beings from the outside but is in fact a creation of those beings. The environment is not an autonomous process but a reflection of the biology of the species. Just as there is no organism without an environment, so there is no environment without an organism. (1983, p. 99)

> It is impossible to avoid the conclusion that organisms construct every aspect of their environment themselves. They are not the passive objects of external forces, but the creators and modulators of those forces. (1983, p. 104)

> In this sense the environment of organisms is coded in their DNA and we find ourselves in a kind of reverse Lamarckian position. Whereas Lamarck supposed that changes in the external world would cause changes in the internal structure, we see that the reverse is true. An organism's genes, to the extent that they influence what the organism does in its behavior, physiology and morphology, are at the same time helping to construct an environment. So, if genes change in evolution, the environment of the organism will change too. (1991, p. 112)

Lewontin's view, as expressed above, will be called a "strongly constructivist" conception of the environment (though Lewontin's view is far from the strongest possible version of this idea). So merely accepting that there are "constructive" explanations in the sense of this book, does not commit one to constructivism in the sense defended by Lewontin.

One mark of strongly constructivist views is the idea that internal organic changes imply environmental changes. I will argue against this component of strongly constructivist positions.

We should note that Dewey in several places explicitly accepts a view of environments which resembles Lewontin's.

> [W]ith every differentiation of structure the environment expands. For a new organ provides a new way of interacting in which things in the world

that were previously indifferent enter into life-functions. (1938, p. 32; see also 1929b p. 179)

There is, of course, a natural world that exists independently of the organism, but this world is *environment* only as it enters directly and indirectly into life-functions. (1938, p. 40)

There is, then, a genuine sense in which the evolution of life, the increase in diversity and interdependence of life functions, means an evolution of new environments, just as truly as of new organs. (1911b, p. 438)

I will argue that this is not a good conception of environment and of environmental change to use in the positive project of this book. It can also be argued that Dewey himself would do better not to have this view, as it interferes with some of his central aims. Dewey wants to insist, against people like Spencer, that organic action involves making changes to the environment rather than merely accommodating it. But if internal changes *are* changes made to the organism's environment, then the difference between the two views is obscured; a passive accommodation of the environment entails a change made to the environment.

Negotiating between the competing demands of different things Dewey said is not the main aim of this discussion though. The main aim is to outline a definite view of what organic construction of environments is, for use in a theory of the place of mind in nature.

We will start by considering some of Lewontin's arguments for his strongly constructivist view. I will group the arguments into five categories.[7]

1. Organisms select their environments
Organisms determine which specific regions of their physical surroundings will be their environments. For instance, a tiny insect can live in small pockets of space which have very different physical and chemical characteristics from their surroundings.

2. Organisms determine what is relevant
The structure and behavior of an organism determine which aspects of its immediate surrounds are relevant aspects of its environment. An area may be polluted for one organism but not for another, according to their different metabolic needs.

3. Organisms alter the external world as they interact with it
An example of this is the creation of the oxygen-rich atmosphere round the earth, which was discussed earlier. There are also more short-term

effects, as organisms consume resources, restructure the landscape, and pollute it.

4. Organisms transform the statistical structure of their environment
Here we assume that some environmental variable is relevant to the organism, and the issue is the environment's pattern with respect to that variable. Several phenomena can be distinguished:

4a: The organism's *size* determines whether a meadow is homogeneous in temperature and light intensity, or a spatial patchwork of hot, cold, light, and dark. Size affects whether the environment is variable or not.

4b: The organism's *longevity* and *mobility* determine the pattern of variation. An immobile or short-lived organism encounters a patchy environment as an uncertain one; it will find itself dealing with one environmental condition or the other. A mobile or longer-lived one faces such patterning as a sequence of states that may have predictable statistical properties.

4c: An organism's capacity for *buffering* and *storage* determine whether resources are effectively variable or not. If a spider needs only to eat once a month, variation in food supply over days is not encountered as real uncertainty. For a small bird which must eat constantly, such variation is of great importance.

5. Organisms change the physical nature of signals that come to them from the external world
Any environmental effect on an organism is filtered and transformed by that organism's intrinsic properties. An environmental change that begins as a change in the rate of vibration of air molecules is transformed by the body into changes in the concentration of certain chemicals, which are perceived by the rest of the body.

Lewontin takes these five families of arguments to all point in the same direction, towards a need to replace a view of life based on adaptation with a view based upon construction. I will argue that these arguments make use of two different sorts of considerations; the list includes both internalist arguments and constructivist arguments. Lewontin's "constructivism" makes use of two different senses of "construct" – a literal causal sense, and a constitutive or ontological sense. The view I will defend is based upon a strict separation between these senses. The constructive role of intelligence, and of organic activity generally, that is endorsed in my position, involves only the causal sense of construction.

5.4   Varieties of construction

Let us next look more closely at the distinction between two senses of the term "construct" that are relevant in this domain. First, there is the type of case where an organism constructs some aspect of the world in the everyday sense in which to build a concrete car park or dam a river is to construct an environment. This is a matter of physical, causal intervention in the world, intervention which effects a change in external affairs. The other sense in which organisms can be said to construct their environments asserts not a causal but a constitutive or ontological dependence. Features of the environment which were not physically put there by the organism are nonetheless dependent upon the organism's faculties for their existence, individual identity or structure.

With this distinction in mind, let us look at each of Lewontin's arguments. The two clear cases are arguments 2 and 3. Argument 3, *Organisms alter the external world as they interact with it*, is an argument based on the causal impact organisms have upon the world. This is literal construction of the environment. Argument 2, *Organisms determine what is relevant*, is different. This argument appeals to a distinction between environmental conditions that make a difference to the organism's life and conditions that do not. If this is to be understood as "construction" of the environment, it is not construction in a sense involving causal impact on the world. The same is true of arguments 4a and 4c. These are arguments which relate the statistical structure of the environment to the organism's point of view. Argument 5, which involves the transformation of signals from the world, is in this category also.

The other cases are more complicated. Argument 1 is based upon the behavioral selection of favorable environments by organisms. This is not a physical change to intrinsic properties of the world, as the creation of oxygen was. But the organism does change its physical *relation* to the environment by moving through the world. Effects on the statistical structure of the environment due to the organism's longevity and mobility, which figure in 4b, are similar. Though these cases are not as simple as argument 2 in this respect, it is still true that if they are to be regarded as "construction" of the environment this must be understood in a non-causal sense. So argument 3 is distinct from all the others in this respect. And it is construction *in the sense of argument 3* that should be embraced as part of, or consonant with, the environmental complexity thesis.

Organic construction of the environment exists whenever an organism intervenes in formerly autonomous physical processes in the external world, changing their course and upshot. This is something some organisms do more than others; there is a difference between termites, beavers and people, on the one hand, which physically shape the landscape they occupy, and cockroaches, hawks and dolphins, on the other, which have much smaller effects on their surrounds. There is a difference between dealing with the world by intervention in it, and dealing with the world by effecting an internal change; a difference between adapting to an irregular terrain by acquiring better balance and more nimble feet, and adapting to an irregular terrain by laying an acre of concrete on it. From my standpoint one of these, the latter option, is a case of organic construction of the world and the former is not. The former course is one of internally adjusting to circumstances. On the strongly constructivist view, construction can be achieved either by physically removing some feature, or by altering the organism's faculties and behavior so the feature is no longer relevant.

When I say there is a difference between intervention in the processes of the world and internal adjustment to them, I do not mean to imply that one mode of adaptation is more "advanced" than the other. It is quite common to think that one concept of evolutionary progress or directionality that makes sense is a trend towards increased intervention in external physical events as a way of dealing with the environment (Dewey, 1917, p. 62).[8] This is an idea to be cautious with. If understood in the most straightforward way, this is a criterion on which worms and termites exhibit a more advanced or developed grade of adaptation than hawks and dolphins. I do not want to put down the worms, but it is certainly true that some environments provide far more opportunities for active intervention in their processes than others do, and some ways of life make this intervention more useful than other ways do. So I will not adopt a "directional" interpretation of the relation between construction and other forms of organic response. The argument is just that these are two different types of relation between organism and environment.

The constructive activities of organisms that I have been taking as paradigms involve organic actions that make changes to intrinsic properties of environmental features. The actions of worms on soil and beavers on streams are of this type. We should recognize, in addition, organic actions that make alterations to relational properties of environmental features. But a proviso accompanies this inclusion of relational properties: organic construction of the environment includes changes made to relational properties as long as *some* change is made to an intrinsic property of something external to the organic system.

One of the activities organisms can engage in is restructuring their environments by bringing certain things closer together, separating others, altering the structural relationships between different kinds of entities. But if the proviso about changes to intrinsic properties is not included, then we arrive again at a view in which internal changes that simply accommodate the external are viewed as construction. Suppose an organism develops a way to detoxify some chemical in its environment which was formerly highly poisonous to it. This organism has made an internal change to its chemistry, and it has also made a change to the relational properties of the external chemical. There is now one less thing the chemical can poison. But if the organism has not, in doing this, made any change to an intrinsic property of any external feature then this is a paradigm case of an internal accommodation of the environment. It is the type of thing to be *contrasted* with constructive actions such as physically removing the chemical from the environment or spraying something on it to change its intrinsic nature.

This is a case in which the proviso disqualified a change where the relational property in question was a relation between the environmental feature and the organic system itself. There are also cases excluded by the proviso in which an internal change made by the organism alters the relations between two external things. Suppose a bird acquires a new skill – using stones as anvils on which to break snail shells. This fact about the bird's structure and dispositions brings it about that there is a certain new relation between these two external things, stones and snails. This change in many cases will lead to changes made to intrinsic properties of the environment. It might lead to snails suffering more predation in areas where there are stones around than in areas where there are not. But until there are changes made to intrinsic properties of things other than the organism, there is no construction of environment.

Saying this is not saying that this relation between stones and snails is less *real* than others. A bird is a piece of nature, and a causal channel running from stones to snails *via* the bird is as real a relation, and as real a causal channel, as any other. This channel can be recognized as a real relation linking snails and stones even before the behaviors of the birds have made changes to the intrinsic properties of any of the local snails. When an organic system makes an internal change that implies a certain change to the relational properties of some external thing, the result is as real a relational property as any other. "Being poisonous to X" is as good a property as "being to the south of X." Our concern is with the question: what is it for an organic system to change the conditions in its environment? There is a difference between organic actions that make changes to intrinsic properties of external things and organic actions that do not, and a useful theoretical framework should not obscure this difference. The present investigation aims, for instance, at formulating a framework within which researchers can ask and answer the question: when and to what extent do organisms construct their environments rather than simply accommodate them? If all changes made to relational properties are counted as construction of the environment then this question is trivialized. Any change that an organism could possibly make in response to its environment would count as construction.

With this narrow category of construction outlined, in the next section some more positive claims will be made about the status of different types of environmental properties. But before this there is a final point to make about Lewontin's arguments. I have run through them with the purpose of distinguishing one out of the range of phenomena he discusses, and giving it a special status as a constructivist argument. If this distinction is made, is it nonetheless possible to recognize a common thrust in all of Lewontin's arguments 1–5? Is there a single message they can all be seen as embodying?

There is a common theme which can be recognized within the present framework. All Lewontin's arguments are arguments against asymmetric externalism in evolutionary biology. Asymmetric externalism can fail through the failure of the environment to be physically autonomous, and it can also fail as a consequence of the organism so transforming environmental inputs that the intrinsic nature of these inputs is of little value in explaining what the system does. Several types of activity – filtering inputs, and physically changing the world – can produce the failure of the same explanatory strategy.

5.5   What environments contain

Suppose we accept that the only construction engaged in by organic systems is the type exemplified by the creation of atmospheric oxygen, the damming of streams and the shading of the ground by tall trees. When an organism is large enough to ignore light differences on the scale of inches, or mobile enough to stay in the shaded parts of the rocks, or nocturnal, this does not count as construction. But then what do we say about the role which organisms play in determining which properties of their environments are relevant to them?

We should say that and no more: *relevance* is a good concept to capture these phenomena. Not every aspect of an environment is relevant to a given organism, and each organism determines by its own size, physiology and behavior which parts of its environment are relevant to it. An organism's environment is its surroundings, and that part of the physical world will have a certain intrinsic structure. In virtue of that structure, any organism's surroundings will have a large range of properties, only some of which will be relevant to it. If a region fluctuates a moderate amount in temperature from hour to hour, this will be relevant to some organisms and irrelevant to others. Organisms *make it the case* that this fluctuation in temperature is relevant or irrelevant to them; because of an organism's size, metabolism and so on, this fluctuation either makes a difference to its life or does not. But to make it the case that some thing is relevant is not to make that *thing*, the thing that is relevant. Being small makes particular types of fluctuation in temperature relevant, without making those fluctuations in temperature.

A mobile organism has the capacity to make different parts of the world into its surroundings. As an organism's surroundings constitute its environment, by migrating to a new place an organism can bring it about that a different part of the world constitutes its environment. This is not recognized in my view as construction of the environment. It is making it the case that a certain piece of the world is environment, but not making or changing that piece of the world. (Given this fact about mobility, the expression "making a change to the environment" is an ambiguous one on my view, and an expression I will avoid.)

In Chapter 4 a distinction between proximal and distal environmental conditions was used in formulating a Dewey-inspired description of environmental conditions in which cognition is useful. Cognition is most useful when there is distal variability which is relevant to the organism, and where there are stable correlations between these distal conditions and proximal, observable conditions. "Relevance" and "observability" have a similar status, on the position defended here. An organism makes it the case that an objective feature of the world is relevant or irrelevant to it, and also makes it the case that a feature is observable or unobservable. This is a consequence of the organism's structure and capacities, and a relevant or observable condition need not be actually encountered or observed. The organism also makes it the case that a condition is proximal or distal, but does this by its physical and spatio-temporal relations to that property.

This is a "realist" view of environments; environments are recognized as possessing objective structure. By virtue of their structure they possess an indefinitely large range of properties that will only be relevant to specific types of organisms, properties which are real features of the environment nonetheless. There is no single description of an environment which is privileged in the abstract. Different categorizations and descriptions of the real structure will be relevant in different circumstances. The structure of the environment will make true a variety of statements making use of different categorizations of conditions. Once a categorization is specified, statements about what the environment contains will be true or false objectively.

Some of these properties will not be obvious, familiar ones. Many parts of Australia, it turns out, are friendly-to-rabbits. They were friendly-to-rabbits even before the rabbits arrived; this property is instantiated, or not, independently of the actual distribution of rabbits in the world. When a new organism arrives in an environment, properties which had not previously been important can become important. When rabbits arrived it was suddenly a very important fact about Australia that it was friendly-to-rabbits. The rabbits did not *make* Australia friendly-to-rabbits (at least, not just by virtue of arriving). But once they arrived, the country's rabbit-friendliness began to have a significant effect on Australia's ecology.

A property like friendliness-to-rabbits may be understood as a second-order property – a property of having *some* structural property such that if certain conditions occur then certain consequences will follow. That is, this property can be understood much in the way that some philosophers understand colors. Redness (or redness-for-humans) can be viewed as the property of having some structural property such that if a certain type of observer is placed in relation to the object in a certain way, a specific type of sensation will result. Whether or not this is the right way to understand redness (where there may be special problems involving the privacy of sensation) it is possible to understand some features of the world in this way, and it is the right way to think of rabbit-friendliness as a property of environments.

A property like rabbit-friendliness will be instantiated or not by an environment independently of the activities and distribution of rabbits. There is no constitutive dependence between what actual rabbits do and an environment's rabbit-friendliness. But there are other environmental properties that are dependent on the activities of actual organisms in this way.

We have already noted that whether or not a given region constitutes an organism's environment depends on the actual location of the organism. This is not a dispositional property. In addition, when an individual organism establishes a territory for itself, it makes a portion of the physical world into *its* territory, and this property can be important in various biological explanations, as many animals behave differently when they are inside and outside their territories. The property of being an individual's territory is not a property like rabbit-friendliness. It does not exist in the environment until the animal in question makes the region into its territory. The same is true with the property of being a mate of a particular individual – this involves an actual causal relationship to the individual in question, not just dispositions or location.

150

These properties of *being a mate*, and *being a territory*, have a similar status to various social and legal properties in everyday life. For land to be zoned as residential, or for a person to be a priest, is not just a matter of dispositions. If he has not been ordained in the right way by the right people he is an imposter, and building a house on a block of land does not make it a residential block. Properties such as these require actual causal connections between the person (or block of land) and particular historically located institutions.

In both the biological cases and the social cases, a change made to these relational properties is considered construction of the environment on my view as long as it is a consequence of organic actions that have made a difference to the intrinsic properties of some environmental features. Signing a piece of paper is not altering the intrinsic properties of a distant block of land, but it is making an intrinsic change somewhere. Leaving a scent on the perimeter of an area of land, and interacting in a particular way with a particular individual, can have similarly far-reaching consequences. But as before, when no change is made to any intrinsic properties of external things this is not construction of the environment.

5.6 Other views

Before leaving this general discussion of properties of environments, I will say something about how my position compares with some others. First, we will look at the view of environments developed in detail in recent years by Robert Brandon and his co-workers (Antonovics, Ellstrand and Brandon, 1988; Brandon, 1990; Brandon and Antonovics, forthcoming).

Brandon and his colleagues distinguish three different "concepts" of environment, the *external, ecological* and *selective* environments. The state of the external environment can be described independently of the properties of the organic system. The ecological environment "reflects features of the external environment that affect the organisms' contributions to population growth" (1990, p. 82). Intuitively, this environment contains only things that "matter" to the organism, where what matters is what affects absolute reproductive output. Third, there is the selective environment. This is intended to be the concept relevant for (micro-) evolutionary theory.[9] Specification of the selective environment requires a comparison between two types of organisms within the population we are investigating. The selective environment contains conditions that make a difference to the *relative* reproductive output of the specified types. According to Brandon, for the theoretical purposes of micro-evolutionary theory a heterogeneous environment is one which varies

151

with respect to the *selective* environment. A physical region which varies with respect to a property that makes no difference to the organisms in question would play the evolutionary role of a homogeneous environment. Also, if the region varies with respect to some property but this property affects all the organisms in question in exactly the same way, this would also play the role of a homogeneous environment. A heterogeneous environment is one whose distinctions have the potential to make an evolutionary difference.

In most respects Brandon's account is compatible with mine. He does distinguish, most importantly, between causal and constitutive relations between organism and environment; Brandon and Antonovics (forthcoming) distinguish "synchronic" and "diachronic" dependencies of environment on organism, and these correspond to constitutive and causal relations respectively.

Is there a real difference between my view and theirs? There is at least a verbal difference, which I do not think is trivial. Brandon sees his account as describing three "concepts" of environment. I would say there is just one concept of environment – the surroundings of a given organism or population, including all the contents of this region. Like any other physical entity or region, this environment can be described and measured in a variety of ways. It can be described in ways that involve the reproductive prospects of a given type of organism, for instance. But that is not to say that the "ecological environment" is a distinct entity. Brandon says that the ecological concept of environment "reflects" features of the physical environment that affect the specified organisms' reproductive potential. Does this mean that it only contains these features? If so, it contains different things from what the external environment contains. But then it would be a distinct thing. I think Brandon's real focus is on different criteria for measurement and classification of conditions – he describes external, ecological and selective modes of measurement and classification. These modes of measurement and classification can all be brought to bear on the one thing – the environment of an organism or population – according to different theoretical needs.

What matters most here is the distinction between causal and constitutive dependencies of environment on organism, and the distinction between internalist and constructive arguments, not the language used to mark these distinctions. I outlined Lewontin's view of environments and then developed an alternative view which makes very little of constitutive dependencies, and which only uses the term "construct" in a narrow causal sense. I do not want to suggest that mine is the only view which can make the right distinctions.

Russell Gray (1992), for example, accepts Lewontin's arguments for the mutual dependence of organism on environment in a more wholesale way, but he also distinguishes between ways in which organism and environment are "co-defining" and ways in which they are "co-constructing" (pp. 176–77). "Construction," as Gray presents it, appears to involve changes made to intrinsic properties of external things. The dependence of environment on organism described as "co-definition" is a constitutive dependence and does not require changes to intrinsic properties of external things.

More generally, now that we have outlined and applied a concept of "relevance" in this context, the view of environments held by Lewontin himself might be expressed by saying that the environment is the *set of relevant external conditions*. If so, what is then required is just a sharp distinction between making a change to the *set* of conditions that are relevant to the organism's life, and making a change to some of those conditions themselves. So there are ways to accept a view of environments which is closer to Dewey's or Lewontin's than mine is, while making a sharp distinction between causal and constitutive/ontological dependencies. The distinction is more important than the precise way in which it is made.

5.7   The status of complexity

An environmental property which we must pay special attention to here is complexity. Earlier chapters discussed explanations of organic complexity in terms of environmental complexity. Complexity was, in both cases, understood primarily as heterogeneity; a complex environment is variable or heterogeneous. But to what extent is complexity a real or objective feature of environments? Lewontin says that the statistical structure of the environment is constructed by the organism. Even if we are determined to resist this language in general, what sense can be made of an objectively complex or simple environment?

There is no need to make a special provision for complexity properties here. Complexity properties are real features of environments which exist independently of organisms. If an organism is to construct or transform the complexity in its environment it must do this by physical intervention in it.

However, environments can be variable in many different *respects*, and only some types of variability will be relevant to any given organism. There is no need for an advocate of the environmental complexity thesis to say that there is a single property of complexity, on which all environments can be compared and ranked once and for all. There are different types of complexity, and some will almost always be irrelevant to real organisms. The important point is that *irrelevant complexity is still real.*

Specification of a complexity property always involves at least specification of (i) the categories or classification of states utilized, and (ii) the measure of variability used. With respect to (i): to say an environment is complex, or more complex than another, is always to say it is complex with respect to certain properties, and the distinctions made between different states and conditions may be more or less fine-grained, for example. With respect to (ii): sometimes the relevant measure will be the sheer range of the different states which occur, sometimes it will be a variance measure, and sometimes perhaps a measure of entropy. In some cases the complexity of the process generating successive states of the world might be relevant (such as the "order" of a Markov chain).

Once there is specification of the relevant measure and the way states are to be distinguished, the level of measured complexity is determined by the actual pattern of the environment with respect to that classification of states. It is no part of a realist position, and no necessary part of the environmental complexity thesis, to say there is a unique measure of complexity. An environment has its own intrinsic structure, but one and the same environment can, by virtue of its structure, be simple in one respect and complex in another.

At this point it is possible to make use of the framework developed in this chapter to present and then respond to an important internalist attack on the environmental complexity thesis.

It has been conceded that any given environment will be simple in some respects and complex in others. The sea is homogeneous with respect to properties of light and temperature, but has a very diverse collection of species. Is this a simple environment or a complex one? One climate might be predictable from day to day but undergo large fluctuations across seasons. Another might be less seasonal but very variable from day to day. Which is the more complex? It was also conceded that an organic system, by virtue of its internal properties, will make some kinds of complexity relevant to it and others irrelevant. The organism plays a role in making it the case that its environment contains *relevant complexity* or not. But what is then left of the "environmental complexity thesis"? It is now apparently up to the organism whether the environment it encounters is one that contains the type of complexity which makes cognition useful or not.

If an extreme illustration of this problem is desired, we can make use of Goodman's (1955) problem with the predicate "grue." Something is grue if it is either first observed before 2500AD and green, or if not first observed before

2500AD and is blue. (This is a *disjunctive* condition; an individual thing does not have to change color to be grue.) Is a world in which all emeralds are grue more complex or more simple, with respect to emerald-properties, than a world in which all emeralds are green? It depends on what is counted as being in a different state. This example seems to suggest that any environment at all can be complex or simple to any degree at all, given an appropriate categorization of states.

Some people will think that this case involving grue is not a problem; one way to deal with arguments like Goodman's is to claim that grue and its ilk are not real properties. Then complexity and simplicity with respect to grue-like properties will not be real features of environments. Whether Goodman's extreme example is accepted or not, the problem he is illustrating is a real one, and one that can be presented in the form of a general challenge to externalist views. Externalists want to explain various organic properties in terms of an adaptive response to specific environmental patterns. But patterns are ubiquitous and cheap. Only by virtue of internal properties are some patterns relevant or salient and others irrelevant. Thus – the objection runs – external properties of pattern cannot be explanatory of what an organic system is or does.

In the face of this argument, the view of externalism that I favor can be expressed with a *concession* and with a *bet*. The concession to be made is that the organic system in question does play a role in determining whether or not a given environmental pattern is relevant to it. There is no way around that. But the properties of the organic system that make the environmental pattern relevant need not be the same organic properties that the environmental pattern can help to explain. The externalist bets that once one general set of organic properties have played their role in determining that some environmental patterns are relevant and others are not, there will be other organic properties that can be explained in terms of this environmental pattern.

Note that this is an internalist concession, not a concession to a strong form of constructivism. The organism, by virtue of one set of organic properties, makes it the case that a given environmental pattern is relevant. It does not construct that environmental pattern.

That is the externalist's bet. One type of internalist can be seen as making a contrary bet. This internalist bets that once we have said enough about the organic system to determine that a given environmental pattern is relevant to it, there is little or nothing left to explain in terms of that environmental pattern.

An example should make this clearer. By virtue of a set of very general internal facts about everyone reading this book, variation in temperature of the type found across seasons in the north-east of the United States is relevant, in principle, to them. Sometimes it is very hot there and at other times it is very cold. This is not to say that everyone actually has to contend with this variation in their lives, but everyone is an instance of the type of organic system that is affected by this type of variation. Because a certain type of variation is relevant to us, and the north-east of the U.S. really does exemplify this pattern of variation, the north-east of the U.S. is a relevantly complex environment, with respect to temperature, for creatures like us. Temperature, consequently, is worth *tracking* in Boston. If you are travelling there it is rational to attend to what season it is and pack your bags accordingly (given anything but a very self-destructive set of goals). If there was this type of variation in San Diego, California, it would be relevant to people who go there. But there is not. In this respect San Diego is a simple environment. If you are travelling to San Diego you do not need to pay nearly as much attention to the season.

So a set of general properties about human beings make it the case that a certain type of variation in temperature is relevant to them. But, the externalist says, there are many other properties of human beings that can be explained in terms of that environmental pattern. For example, the different behaviors of travellers to Boston and to San Diego can be explained in terms of the different properties of pattern in those two environments. The more complex environment generates more complexity, of a certain type, in the behavior of travellers.

Here, as elsewhere, in assessing the merits of externalist views we need to attend carefully to the explanandum involved. If the question is why people are sensitive to temperature variation of the type found in Boston, the answer is given in terms of a set of internal properties of human beings. If the question is why travellers to Boston take heed of the season when making their plans, the answer is given in terms of Boston's real properties with respect to climate.

An externalist can in many cases be viewed as someone who thinks this sort of case is typical: there is a gap between the organic properties which make a pattern relevant and the organic properties which the pattern might explain. Internalists of certain kinds might be viewed as betting that this case is not typical.

This is the reply to the challenge made with the extreme example of grue-like properties as well as more familiar ones. Variation in temperature is a real feature of some environments, and some properties of organic systems are responses to this type of variation. Variation in temperature* (a grue-like derivative of temperature) can be admitted to be a feature of certain environments as well. But it is an idle feature of the world. The fact that temperature and not temperature* is important in causing certain properties of organic systems is not compromised by our not being able to metaphysically rule out temperature* in advance.

In principle there might be an environment whose *only* properties of heterogeneity were grue-like. Matter and energy might be distributed in a way that is, by our usual measures, completely uniform and stable. Is this a complex environment? Here we should distinguish between what the project of this book requires, and what might be shown with further metaphysical argument. In the official view of this book, such an environment has a lot of heterogeneity properties, but none that will be relevant to most organic systems. It might be possible to argue for a stronger view, and exclude as illusory or artificial certain types of heterogeneity (Lewis, 1983). Then we could say that this environment has no genuine complexity properties at all. It is just *simple*, and that is the end of the story. This is certainly an attractive view, but it is not required here, and it would require further argument to support it.

This marks another point on which my view diverges from the view of someone like Spencer. Spencer, as far as I can tell, did think that there are complex environments and simple environments. The sea is simple and the land is complex, and on land the tropics present "less varied requirements" for organisms than the temperate zone (1866, pp. 85–87). This is not an assumption required in the present view. Cognition is an adaptation to *properties* of environmental complexity, properties that are found in different specific forms in different environments. Cognition is not an adaptation to "the" complex environments.

5.8 Construction and realism

Some readers may wonder why I have been so insistent that "construction" be understood in a certain way, and so concerned to distinguish sharply between different types of dependence between organisms and environments. Is the issue this critical?

Part of the motivation for this close attention is the importance which the concept of construction, and constructivist views, have acquired within philosophy in general. The perennial debate between realists about the external world and various types of antirealists is now often carried on in the language of construction. An attitude of "social constructivism" is also powerful or dominant in many fields in the humanities and social sciences. Some find this view to be revolutionary and emancipatory. Others find it a plague.

The view defended in this chapter is based upon several distinctions, which are understood as sharply as possible. One is the distinction between an environment's own structure and the different possible ways of describing that structure. Another is the distinction between literal and metaphorical senses of "construct," and a third is the distinction between an organism's "perspective" on the world, and its actual causal effects on the world. These distinctions are characteristic of a generally "realist" outlook.

Some constructivist positions can be understood as denying the importance or the coherence of the distinctions upon which my view is based. A constructivist might claim, not that the two senses of "construct" mean exactly the same thing, but rather that both sets of phenomena – our physical action on the world, and the operation of our categories and languages – have a similarity which makes the term "construct" appropriate in both cases. Constructivists think that realists implicitly downgrade the efficacy of human action, and also the multiplicity of possible interpretations of phenomena. Some think that admitting a world which exists autonomously, independent of our thoughts or social practices, encourages an attitude of subservience to external things, as opposed to a view which places human creativity and possibility in center-stage.

The discussion in this chapter is intended to have relevance to these broader debates. It is intended, first, to outline explicitly a picture of a causal structure which some constructivists might recognize as their real foe: asymmetric externalism. This is a *causal structure*, not a metaphysical position. Second, this discussion is intended to motivate recognition of the difference between internalist and constructivist lines of dissent from this picture. Thus it is hoped to remove some of the temptation towards over-strong constructivist positions. An environment's *effects* on an organism may be heavily dependent upon the organism's own properties; the organism may engage in massive elaboration, transformation and amplification. The external stimulus may play the role of a mere fuel or impoverished trigger, giving rise to the most extravagant flights from within. But rejecting external*ism* can be distinguished from de-externalizing the external, in the fashion of strong forms of constructivism.

The distinctions that my view is based on can be outlined and illustrated, but it is another thing to convince someone that they provide the best framework for thinking about the relations between internal and external. In this section I will give a more positive argument for holding to a narrow construal of "construction," for the purposes of theorizing about relations between thought and the world. This argument is derived from Dewey.

On construction

Dewey's view of environments is not the same as the one defended in this book, as noted earlier. But aspects of Dewey's position on the constructive role of thought, and its effects in the world, are consistent with the present framework and provide a powerful argument against certain modern forms of constructivism.

Dewey was opposed to asymmetric externalism as a picture of our relations with nature. He was opposed to views which obscure the scope and potential of human initiative in the world, views which artificially magnify the power and permanence of external forces. He believed that thought and theory have a role in the construction of the world. But he did not mean this in the way it is meant by "constructivists" today. Because for Dewey, the only way to transform the world is to intervene in formally autonomous processes in nature and alter their course and upshot. And thought only exercises this role insofar as it is expressed in action.

This was the core of Dewey's argument against the "idealism" of his day in *Experience and Nature*. Dewey agreed with idealists that thought is a real force in the world. But he opposed the idealist view that thought can effect changes to the world without this influence going via action. Idealism mistakes the real role of intelligence in aiding efforts to reconstruct our environments, for thought's having a role in the original construction of the world.

> [I]t is not thought as idealism defines thought which exercises the reconstructive function. Only action, interaction, can change or remake objects. The analogy of the skilled artist still holds. His intelligence is a factor in forming new objects which mark a fulfillment. But this is because intelligence is incarnate in overt action, using things as means to affect other things. (1929a, p. 158)

For Dewey, metaphysical idealism is an attempt to hang onto two ideas at once, one true idea and one false one. The true idea is that thought affects the world. The false idea is that thought should be understood independently of its role in directing action. This combination is what results in the view that mind, or language, or theory, in isolation of action, can make the world what it is, can change it or construct it. This is what results in the idea that " 'mind' constructs the known object, not in any observable way, or by means of practical overt acts having a temporal quality, but by some occult internal operation" (1929b, p. 19).

There is one significant difference between Dewey and me here. Suppose a new theory of some domain makes possible new forms of action. The things described in the theory acquire new potentialities as a consequence, even

159

before action has taken place. They acquire new possibilities for interaction with us and with other things. Dewey does regard this as a change made by the new theory to its subject matter. I am holding out for a change made to intrinsic properties, not just potentialities, before thought has changed the external world. But Dewey's position here does not compromise his insistence on action; the new potentialities only exist because of the new scope for action (See Dewey, 1922, p. 298; 1929a, p. 157.)

Dewey's criticism of idealism applies also to much constructivist thought today. There is a definite domain in which humans can reconstruct and transform what they find, and within this domain intelligence and creative thought are vital to the task. But there is *only* a certain domain that is plastic in this way. We can firstly construct concepts, patterns of thought and practice. We can construct and engage in new forms of social interaction. We can also use these products to transform and reconstruct aspects of nature external to ourselves, via action. But only some parts of nature are susceptible to this treatment. There is no way that the dinosaurs, the stars and much else can be constructed in this way. We can change their significance in our lives, and change various of their relational properties to ourselves and to other things existing in the present. But the only sense in which the stars themselves could be constructed by us is via "some occult internal operation." And within the domain in which this constructive activity on nature is possible, it is not achieved by the mere deployment of categories or a creative internal leap. These are ingredients of the process, but only ingredients. A basic mistake of metaphysical constructivism is to think that representation, or categorization, or creativity, still has the power to construct the world when it is considered in isolation of the mechanisms that physically effect the construction. Another common error is the mistaking of a genuine asymmetric externalist enemy for an imagined realist one.[10]

Constructivists are not the only people Dewey's argument can be directed against here. In this book Spencer has received a comparatively generous hearing. But perhaps the single central idea in all Spencer's work falls prey to a criticism of Dewey's which is part of the same line of thought as this argument against "idealism." It was central to Spencer's work to assert the existence of processes in the world which will inevitably tend towards progress. Progress for Spencer is a "beneficent necessity" (1857, p. 60). Dewey sees ideas of this type as attempts to assert the reality of good in the world without recognizing the need for people to achieve it by specific practical measures. It is trying to give a metaphysical guarantee for something that can never be guaranteed, that can only be gained in a step-by-step and fallible way, by bringing intelligence and effort to bear on problems. To assert that mind can construct the world on its own, or to say that the universe is guaranteed to take the right path without being made to, is in either case to downgrade the real constructive role of intelligence.

5.9 Constructivist philosophies of science

I will conclude this part of the discussion by looking again at an example discussed in Chapter 2, which is intended to support my claim that the function of strong constructivist claims is often to oppose asymmetric externalist views.

Empiricism, as outlined earlier, is in some versions an externalist explanatory program in the philosophy of science. Empiricism will often be an asymmetrically externalist view here, also. Whether the subject matter of science is an autonomous external realm, or mere patterns in the "given," in either case the scientist is conceived as responding to these facts without substantially altering them.

Many have thought, especially since Kuhn (1962), that this picture is a distortion of the nature of scientific practice, that scientific communities have an intrinsic structure which plays a large role in determining the effect of any event at the observational periphery. Some theorists replaced observation with another external factor, the political and social interests of the group or class to which the scientists belong. As outlined in Chapter 2, this led to the use of a set of terms which have muddied the waters. Observation was taken to be a factor "internal" to science, and the interests of a social class were seen as "external." But both empiricism and macro-sociological "interest theory" are externalist programs. (I am not sure to what extent the sociological approaches should be viewed as asymmetric.)

There has also arisen a more internalist view in the field of "science studies" – in this respect the heir of Kuhn. This program is exemplified by Latour and Woolgar's *Laboratory Life* (1986). Latour and Woolgar investigated the internal dynamic of a scientific field without making use of any external directing force, political or experimental. A scientific "fact" becomes solidified and established as a consequence of the activities of workers who restructure networks of scientific belief and practice in such a way that it is too costly to change what is done and challenge the patterns of assumption that are made.

All of these opponents of empiricism – Kuhn, "interest theory" sociologists, and Latour and Woolgar – are associated with constructivism to various degrees, and constructivist attitudes are common in "science studies" in general. Some of this talk of construction can be seen as making an internalist point about the creative activity of scientists. Sociologists of science oppose the idea that reality acts on the scientist with an "unmediated compulsory force" (Shapin, 1982, p. 163). The aim is to view theories as the products of human agents rather than as structures imposed upon them by nature. But at least part of the constructivism in recent history, philosophy and sociology of science is intended to have ontological bite. The facts, or the world, are products of scientific activity: "The argument is not just that social networks mediate between the object and observational work done by participants. Rather, the social network constitutes the object (or lack of it)" (Woolgar, 1988, p. 65).

This has generated a heated philosophical discussion, centered primarily on the coherence of the idea of a world existing independently of all conceptual schemes, and the coherence of constructivist alternatives (Devitt, 1991a). In my view, the commitment to constructivism on the part of radical science studies functions largely as a way of making the most definite break possible with the externalist explanatory program of empiricism. Empiricists wanted to see science as controlled by observation, a channel that conducts information mainly in a single direction: outside-in. One way to deny such a claim is to sever the two domains, to claim that neither influences the other very much. But a more striking way is to reverse the explanatory arrow, as radical science studies did. The external world does not control the course of science; instead it is controlled *by* the course of science. As Woolgar continues the passage quoted above: "The implication for our main argument is the inversion of the presumed relationship between representation and object; the representation gives rise to the object" (p. 65).

Note that it is not externalism in general that is denied by the constructivist move – some sociology of science is both externalist and also constructivist about the external world. The reaction is directed against a particular externalist view, and it works by inverting a particular arrow of explanatory relevance.

It is possible that an underlying premise in some constructivist positions is that a *decoupled* view is not an option: either the world controls representation or representation controls the world. But a decoupled view *is* an option. Perhaps scientific belief and the rest of the world are largely disconnected. This view is even an option within a representational analysis of scientific language. In such a view, both truth and error are real but accidental.

Here as elsewhere, debates which are overtly metaphysical would make more progress if they were focused on causal or dynamic questions, questions about the lines of influence and control. The language of the protagonists, even when waxing metaphysical, often reveals this. That the real issue at hand is the *locus of control* is suggested, for instance, by the definition of reality proposed by Latour and Woolgar in *Laboratory Life*. They define reality as "that which cannot be changed at will" (1986, p. 260).[11] This is among the worst possible definitions of reality, from the standpoint of the present book. A piece of rubber hose and an iron bar have their own real shapes. One can be changed at will and the other cannot. The fact that one cannot be changed at will may thwart us in some purposes, but may be an instrument for others. It can act as a lever, or a stay around which the rubber hose is bent. In any case, problems with this definition itself are less important here than what it shows. Realism is associated by Latour and Woolgar not with the objective existence of some realm, but with its *autonomy*, with its resistance to our efforts. Given this, it is comprehensible both why anti-externalist arguments should be thought to have ontological import, and why realism should be seen, by many modern construc-tivists, as a view hostile to creativity and intervention. But the realist view defended here has none of these associations. The issue of what entities nature contains is one thing; the lines of control which exist between different natural systems is another. Intelligence, creativity and the active reorganization of experience are vital tools for agents, but nothing is constructed, except for the theories and thoughts themselves, until this intelligence is expressed in action.

Extreme constructivism provides a dramatic way of casting aside empiri-cism's externalist explanatory arrow. But the pitfalls of this move must be appreciated. One result is an obscuring of the relation between thought and action which, Dewey argued, produces a disregard for the role of intelligence in practical life. The answer is to assess asymmetric externalist views within a framework that does not compromise realism. Dewey's point in *Experience and Nature* applies again. Science can alter the world by making material changes to it, via the interface of technology. But it is not scientific belief as constructivism conceives of scientific belief, which exercises this reconstructive function. Only action, interaction, can change or remake objects. A paradigm or conceptual scheme can be a factor in forming new objects which mark a fulfillment. But this is because intelligence is incarnate in overt action, using things as means to affect other things.

Notes

1 There could hardly be a more over-used term than "interactionism," but it is the most convenient one and it is used only in a specific sense here. This is not the

"interactionism" of people like Oyama (1985; see section 2.7 above), who treat the statistical sense of "interaction" as providing a model for organism/environment relations. That "interactionist" view is concerned with explaining organic properties, and it claims that neither internal nor external should be regarded as primary. We might call that view "synchronic" interactionism, in contrast with the "diachronic" interactionism that is the topic of this chapter (I owe this suggestion to Richard Francis). It is not uncommon for people to espouse interactionism in both senses. Lewontin (1974b, 1983) and Oyama (1985) may be examples. But the two positions are strictly independent.

Dewey in some of his later works came to use "interaction" as a term for a picture of inquiry and organism/environment relations that he viewed as inadequate. He used "transaction" for the view he preferred (Dewey and Bentley, 1949, Chapters 4 and 5). An "interactionist" view in my present sense is not necessarily "interactionist" in Dewey's sense in that work, and it is not "transactional" either.

The synchronic/diachronic distinction made by Brandon and Antonovics (forthcoming) is related to, but not the same as, the distinction made here.

2 As Rasmus Winther suggested to me, a good illustration of this last, asymmetric constructivist option might be found in football games between teams of small children, whose parents line the field and yell instructions to their unfortunate offspring.

3 For a good discussion of J. S. Mill on this score see Heilbroner (1992). Otto Neurath and some other members of the Vienna Circle provide another good example (Galison, 1990), although the views of the Vienna Circle are also not strongly externalist in the way seen in more classical, psychologistic forms of empiricism.

4 Another often-cited example of adaptationist work is the investigation of the effects on plants of heavy metals in the soil around mines (Antonovics, Bradshaw and Turner, 1971). Many plants have apparently evolved a tolerance to normally toxic levels of lead, zinc and so on. This is a botanical analog of industrial melanism (and Antonovics et al. make an explicit comparison to that most famous case of natural selection: p. 30). The phenomenon looks very asymmetrical at first glance, but in fact Antonovics et al. note that the plants "probably play an important part" in physically altering the toxic environment (1971, p. 22).

Antonovics has also written a paper with Robert Brandon explicitly advocating an interactionist view of organism/environment relations (forthcoming).

5 Dewey blamed Spencer for the popularity of a view of evolution and also education as "the molding of pliable and passive organic beings into agreement with fixed and static environing conditions" (Dewey, 1911a, p. 365). In the light of Chapter 3, it should be clear that the first part of this accusation is fair but the second is not: Spencer saw evolution and organic action in general as the molding of pliable and passive organic beings into agreement with *constantly changing* environing conditions. Dewey (1904) gives a better account (and a surprisingly favorable estimation of Spencer).

6 Thayer (1952) is a sympathetic critique of Dewey's theory of inquiry. On two central issues the modifications Thayer suggests to Dewey's theory coincide with views which are accepted here: (1) Not all inquiry transforms the material of the problematic situation, although some does, and (2) accepting the instrumental role of thought in inquiry is not incompatible with regarding these thoughts, as they function in that inquiry, as true or false.

7 For argument 1, see Lewontin (1977a) p. 53; for arguments 2–4 see (1983) pp. 99–101. For argument 5 see (1991) p. 116.

8 See also Dewey (1948, p. 85). Lewontin himself once accepted an idea of this type (1957, p. 407).

9 Micro-evolutionary theory is concerned with evolutionary change within a single breeding population. Macro-evolution involves speciation, extinction and other larger-scale processes.

10 In addition to antirealism driven by a dislike for asymmetric externalism there is antirealism driven by skeptical worries. This is the type of antirealism which is associated with empiricism, and is exemplified by Berkeley, and verificationism. I would not claim that the verificationist route to antirealism can be understood in terms of revolt against asymmetric externalism. See Boyd (1983) for a treatment of the relation between "empiricist" and "constructivist" antirealism in science which complements the present discussion.

11 It has been suggested to me that I am taking Latour and Woolgar's remark about reality more seriously than they intended it. But see also Latour (1987, p. 93).

Dewey's view, as I understand it, is diametrically opposed. When philosophers take the practical turn he recommends, the real "ceases to be something ready-made and final; it becomes that which has to be accepted as the material for change, as the obstruction and the means of certain specific desired changes" (1948, p. 122).

Dewey in one passage does make a Latour-like point about the concept of an object: "Object is, as Basil Gildersleeve said, that which objects, that to which frustration is due." He then goes on: "But it is also the object*ive*; the final and eventual consummation" (1929a, p. 239). The real can, in many cases, be changed at will, but action is required for doing so.

6

The Question of Correspondence

6.1 The division

In the previous chapter we looked at a basic difference between Spencer's and Dewey's versions of the environmental complexity thesis: Dewey was opposed to conceptions of mind which are asymmetrically externalist. He accepted, with others of a generally empiricist temper, that thought is highly responsive to patterns in experience. But unlike Spencer and unlike more orthodox empiricist views, Dewey placed great stress on the role played by thought in the transformation of external affairs, and hence on the future course of experience. On this point I follow Dewey. But though this is, in my view, the most fundamental difference between their conceptions of the place of mind in nature, it is not the most obvious difference between them. The most obvious difference is in their attitudes to the concept of *correspondence*.

Spencer believed that states of mind, by virtue of their role in adapting organisms to particular external conditions, often bear a relation of correspondence to these external conditions. Dewey viewed his account of the role played by the mind in dealing with environmental problems as a *replacement* for the idea that the business of thought is correspondence to the external world. He saw the idea of correspondence as a typical product of a "dualist" view about mind and nature – philosophers first assert a breach or gulf between mind and the world, and then invent magical relations to overcome it. For Dewey, there is plenty of real traffic between agents and their environments; environmental problems prompt agents to intervene in external affairs. But this causal traffic has little to do with correspondence, with a "mirroring" of the outer by the inner. So Dewey's version of the environmental complexity thesis is intended as an alternative to a more traditional view of the place of mind in nature, a view in which mind has the function of corresponding.

166

The question of correspondence

Dewey used both the labels "pragmatist" and "naturalist" to describe his views, although he was cautious with the former term (1931). In recent years naturalists have tended to side more with Spencer on the question of correspondence, while pragmatists have sided with Dewey. The "naturalistic" work I have in mind here is exemplified by writers such as Dretske, Fodor, Millikan, and Papineau (although Quine, who lies at the root of much contemporary naturalism, is not an advocate of correspondence in this sense).[1] Writers such as Dretske and Fodor have developed theories intended to show how it is possible for inner states of physical systems to represent external conditions. They have proposed theories of meaning for certain inner states, where meaning is understood, at least in part, in terms of truth-conditions. The theories claim that if a certain physical relation exists between an inner state of an organic system and a condition in the world, then that inner state represents that condition in the world. An inner state has the content "S" in virtue of physical properties, and when that inner state is tokened, it is true if condition S obtains in the world. If S obtains the organism's states correspond to the way things really are.

It may be possible to argue with this, but I think this naturalistic work is, among other things, an attempt to make sense of correspondence as a relation between thought and the world. Modern naturalists do not usually think of correspondence as a relation of "picturing," and some readers may think that the concept of correspondence has some essential connection to picturing. On the view accepted here, picturing was just one way in which people tried to make sense of the most basic feature of correspondence: signs which correspond represent the world as it really is. The signs do not have to be similar, in some abstract way, to the external affair represented; no "isomorphism" is required. The core of what is required for a sign to correspond to the world is that the sign have a definite truth-condition. (Perhaps it is better in some cases to say "correspondence-condition" or "accuracy-condition": see section 6.4). What is required is that the sign have, as a matter of natural fact, the property of representing the world as being such-and-such a way, and that the world actually be that way. The naturalistic analyses of Dretske, Fodor and others make sense of correspondence by singling out certain situations as situations in which organisms harbor within them states that represent the world as it is. Whenever meaning is understood as a "robust," natural-world property and is also understood in terms of truth conditions, there is the possibility of correspondence truth.[2]

Pragmatists, on the other hand, have continued to side with Dewey on this point. Richard Rorty is a good example (1979, 1982, 1986). Rorty is opposed to the idea of thought as a "mirror" of nature, as Dewey was. He has also singled out several examples of the type of naturalism I will focus on, as constituting attempts to re-establish the bad idea of correspondence. These are Papineau's and Millikan's biological views of representation (Rorty, 1991a), and Hartry Field's proposed marriage of Tarski's meta-logical theory of truth and a causal theory of reference for terms (Field, 1972; Rorty, 1982). Another writer who has attacked the program of explaining correspondence with naturalistic semantic theories, while flying a pragmatist flag, is Stephen Stich (1990). As Stich is also a naturalist he is, like Quine, an exception to the "pro-correspondence" tendency I am alleging in recent naturalism.[3]

In any case, both history and the course of more recent debates support the view that the status of correspondence is a critical issue for the debate between pragmatists and their opponents. But some naturalistic theories give new twists to this question. Dewey, James and those who follow them on the pragmatist side urge us to replace the false goal of correspondence with the real and concrete goals of dealing with the world or transforming it. But many contemporary naturalists in no way deny the importance of dealing with the world. They hold that the way we manage to deal with the world is *via* having states which correspond, by basing our actions on beliefs that represent the world as it really is. Armstrong (1973) and Dretske (1988), borrowing from Frank Ramsey, view beliefs as *inner maps by means of which we steer* (Ramsey, 1927). Correspondence is viewed as a means to practical success, not as a rival goal. We will look closely at this idea in this chapter – it is an important line of objection to some central aspects of pragmatism.

When this objection is mounted to pragmatism, it is not necessary for the naturalist to deny that correspondence truth can also be valued in itself, if that is what an agent wants. It was accepted in Chapter 4 that, like Dewey, we can recognize value in certain properties of thought over and above their practical value. The present point is rather that correspondence may be important *within* the phenomena described by the environmental complexity thesis. It is not inevitably part of an alternative to that thesis.

6.2   Some false dichotomies

We will begin by looking briefly at what Spencer and Dewey said about correspondence, and at the role the concept played in some of the original debates about pragmatism.

In the third chapter I gave a brief account of Spencer's theory of life and mind. The actions of living systems tend to preserve a "balance" between internal and external processes, preventing the matter which makes up the living body from reaching the state of stable equilibrium with the environment that nonliving bodies reach. Spencer describes this informally as the "adjustment of internal relations to external relations" (1855, p. 374). Spencer also introduces another set of terms in an attempt to explain his definition. He describes the characteristic relations between a living system and the environment as relations of "correspondence" or "reference." He says that life can be viewed as a set of internal changes in correspondence with a set of external changes.[4]

Unfortunately, Spencer's attempts to say what this correspondence consists in are obscure.[5] The inadequacies of Spencer's official concept of correspondence provided Dewey with an ideal illustration of the wrong way to think about living systems. For Dewey, the concept of correspondence only has a place in theories that are trying to reunite an artificially separated agent with its environment. Correspondence is a purely epistemological glue with no place in a naturalistic outlook.

Dewey's attack on Spencer's view of life is related to a more general set of debates around that time, concerning truth and the function of thought. We will look at three different arguments which pragmatically minded philosophers used against correspondence. Apart from stage-setting, this part of the discussion is intended to help establish the outlines of a moderate "orthodox naturalist" view of correspondence which will be discussed throughout this chapter.

First we will look at one of the arguments used by William James against correspondence or "copy" views of truth.[6] This argument works by posing the agent with a choice between two epistemic goals. In one view, an agent regards their epistemic choices as part of an integrated set of individual values and practical goals. Alternately he or she can choose to superstitiously seek a mysterious relation of correspondence or copying between thought and the world. James thought that once we view the issue as a choice between copying the world and dealing with it, we will choose the latter.

This is a false dichotomy. There is no need to regard corresponding to the world as an alternative to dealing with it. It can be a *means* to dealing with it. This is part of the point of the analogy between beliefs and maps. When setting out on a trip, one is not forced to choose between having a map that represents the terrain as it is, and having a map that is likely to lead us to where we want to go. Holding other factors constant, the accurate map *is* the one which will get us where we want to go.[7]

The second line of argument is more characteristic of Dewey. Before discussing it, there is a complication to note. Dewey sometimes simply opposed the correspondence view, but sometimes said that "correspondence" is an appropriate term for describing thought, if it is understood in the right way. Correspondence can be viewed as a relation of harmonious fit between different parts of experience – a cognitive part, and a part that the cognition is intended to help the agent deal with (Dewey, 1910a, pp. 5–6). Dewey opposed the view of truth as correspondence when this is interpreted as asserting a relation between something mental and something external to thought and experience. For example, correspondence is not the relation between the event of Julius Caesar's death and our present thoughts about it. The only "correspondence" here for Dewey is a relation between our thoughts on the one hand and our experiential dealings with present and future traces of the past event on the other. When I say, in this chapter, that Dewey was opposed to correspondence, I have in mind a conception of correspondence as a relation between thought or experience and something external to it.[8]

Dewey often claims that the problem with correspondence as a view of the aim of thought is that, if the idea is coherent at all, it implies or suggests that the mind does not change the world but instead passively adapts to it (1929b, pp. 110, 165). Correspondence, in this view, is (at best) a relation one adopts to conditions that one must take heed of but cannot or will not change. In Dewey's view mind does typically intervene in the world, via its guidance of intelligent action.

In this argument two divergences we have discussed between Spencer and Dewey are seen to rest, at least in part, on one idea. In Chapter 5 I contrasted Spencer's asymmetric externalism with Dewey's interactionist view of organism and environment. In the argument against correspondence I have just described, Dewey finds the same fault with his opponents: a denial of the causal efficacy of intelligence, a demand that the mind observe and accommodate but not interfere. Correspondence as an ideal is, for Dewey, another denial of the active role of intelligence in the world.

Dewey's argument here is based on another false dichotomy. There is no need to choose between corresponding to the world and changing it. Correspondence can be a means to action of a type that transforms the world – even action that transforms conditions represented along the way. One way to put the point, which some naturalists (such as Millikan) take very literally, is to say that beliefs are meant to correspond to the world, and *desires* are meant to change it, by directing action in concert with beliefs. Within the general picture of cognition given by the environmental complexity thesis, beliefs, desires, perceptions, hypotheses, plans and actions can all have different component roles. Not every aspect of cognition has to be directly involved in transforming the world all the time.

Dewey's other views are not inevitably hostile to this account of the role of correspondence. Dewey said agents make use of stable aspects of the world in order to deal with the variable aspects. We could say that agents seek to acquire states which correspond to stable aspects of the world in their attempts to make use of them.[9] But there is no reason not to also accept the stronger view: even those very states of the world which are most ripe for change, the ones that most urgently invite intervention, may be best approached by first acquiring an accurate representation of them as they are. "*Know your enemy!*"

A third pragmatist argument against correspondence, used sometimes by James and Dewey, and also used more recently by Rorty and Stich, is that correspondence as a relation between thought and the world necessarily involves some relation of "picturing" or isomorphism between inner and outer. Stich (1990) says that one of the positive reasons people have had for valuing correspondence is the appeal of a set of metaphors – they think true beliefs are like undistorted mirrors of the world, faithful pictures, or accurate maps. But, Stich says, this is just a metaphor. True beliefs are not like mirrors, pictures or maps.

This is one place where heels should be dug in. True beliefs are not pictures, mirrors or maps. But a true belief, in the naturalistic view we will discuss, has the same relation to the world that an accurate map has. *Accuracy* in an external map is the same basic relation as correspondence truth in a belief. A map is accurate if it represents the world as being some way, when it is that way. A belief is true by virtue of the same relation. Wanting truth in a belief is like wanting accuracy in a map: it is wanting a sign that represents the world as it is, because it is thought that these are the signs that it is best to steer by.

The polarization about correspondence between pragmatists and more orthodox views at this point appears artificial. We have heard no argument against the idea that agents seek to harbor states that correspond to external affairs because this is helpful in their attempts to deal with external affairs and perhaps change them. We have heard nothing against the view that beliefs which correspond are as useful as maps that mark swamps where there are swamps and peaks of 10,000 feet where there really are peaks of 10,000 feet. We will call this the "orthodox naturalistic" view of correspondence.

6.3   A fuel for success

In the orthodox naturalistic view, correspondence is a useful property of some kinds of inner states. This view can be incorporated into the environmental complexity thesis: it can be asserted that the function of cognition is to help the agent deal with environmental complexity, and one way that cognition does this is by making it possible for the agent to direct behavior with internal states that represent the world as it is. In the orthodox naturalistic view, correspondence is a resource, or a *general-purpose fuel for success* in dealing with the world.

Correspondence truth, in this view, is a general-purpose fuel for success in something like the way money is. Money is useful for people pursuing a wide range of projects; it is not a special-purpose fuel for success like kerosene. It is often appropriate to try to acquire true beliefs about a domain even before one has worked out what one wants to do in that domain. Acquiring kerosene in advance is less often appropriate. On this view, it is also possible to give a certain type of explanation for the success of behavior: Jake arrived safely *because* he knew the lie of the land. He thought the quicksand was to the right, and that is where it was.

Truth, money and the goodwill of others are resources for success. Something that is not a resource or fuel for success, in this sense, is *likelihood to succeed*, or *success-proneness*. To call someone success-prone is to say that they tend to do well, but not to isolate some definite structural or behavioral property they have that causes this success to come about. At this point I will make some assumptions about explanation. It will be assumed that citing a disposition to behave in some way is not, in general, giving a genuine causal explanation of that behavior when it occurs. If fragility is a disposition to break, or break-proneness, then to say that something broke because it was fragile is not giving a causal explanation. This is not to say that citing fragility when something breaks is saying nothing at all; an event is claimed to be unexceptional, to be part of a pattern. This is explanatory, in a certain sense. But fragility is not a "fuel for breaking" or a producer of breakings. The cause of the breaking will be some specific structural property that *made* the object liable to break. Certain specific structural properties make butterflies fragile, and other structural properties make champagne glasses fragile.

Similarly, a character called "Simon the Likeable" in the television comedy *Get Smart* had a certain set of properties that made him liable to succeed – it was impossible for people not to take a liking to him, and want to do what he asked. This was a fuel for success, the basis for his success-proneness, and a causal explanation for Simon's unusual achievements. Other people have other fuels for success, and (in the orthodox naturalistic view) correspondence truth is one of these.

It is not part of this orthodox naturalistic view to say that truth (or truth plus rationality) will tend to be the *only* explanatory factors when success occurs. Neither is it denied that truth can sometimes lead an individual into disaster in a situation where a false belief would have saved them. Stich (1990) makes use of cases of this type to argue that there is no interesting link between truth and success after all.[10] Harry has the true belief that his plane leaves at nine o'clock, and he catches it, but it crashes. He would have been better off believing the plane left at ten o'clock. Stich concludes from this, not that truth is no better than falsity, but that truth is no better than "truth*," a relation between beliefs and the world which assigns true* to many truths but also assigns true* to the belief that the plane left at ten o'clock.

If this form of argument is good, then it is destructive not just of common sense respect for true belief but of common sense respect for many other things that people take to be fallible guides to success. An accurate map may help you find a certain house – where the occupants turn out to harbor a violent dislike for you. You would have been better with a map that was "accurate*." You look after your health, but if you had been ill that day you would not have gone out to the unfriendly house. You would have been better off with "health*" (unfailing normal health except a bad cold that day). Money and the goodwill of others have the same property; general-purpose fuels for success can exist, and can be rationally sought, without their bestowing guarantees.

There is much to be said for the orthodox naturalistic view of correspondence. It is in accordance with some compelling aspects of common sense, and it enables us to get over some old debates between pragmatists and their opponents. If this view is accepted, then in this regard pragmatism does *not* succeed in using the environmental complexity thesis to unseat established conceptions of epistemic value.

We have not yet addressed all the objections that can be made to the orthodox naturalistic view though. One challenge comes from advocates of the "deflationary" or "minimalist" approach to truth. According to this view, truth is a mere logical device for making certain assertions in an indirect way, and not an objective property of signs (see note 2). Horwich (1990) has argued that the link between truth and success can be accommodated by a deflationary view. Devitt (1991a) has also argued that epistemic properties such as "warrantedness" or "justifiedness" may explain success just as well as correspondence truth. I will not discuss this set of issues here.[11] The threat to the orthodox correspondence view that we will look at derives from some unsolved questions in the philosophy of mind. What *is* it for a physical system to represent to itself some external condition, to represent to itself that *p* obtains? The difficulty is that some attempts to answer this question in naturalistic philosophy are not so much incompatible with the idea that truth leads to practical success – they are in a sense *too* compatible with it. These theories make the connection between truth and success an *ingredient* in what it is to represent the world. More specifically, part of what it is for something to be a belief that *p* is for it to direct action in a way that is successful if *p* obtains. The problem is that explanations of success in terms of truth threaten to become vacuous, or at least much reduced in content. Of the list of naturalists I gave earlier, Millikan (1984) and Papineau (1987) have views with this consequence. The truth of a belief becomes not so much a cause of behavioral success, but more a conceptual consequence of it.

If this view of correspondence is correct it is not a disaster. First, a naturalist who believes in correspondence could admit that familiar explanations of success in terms of truth do not say as much as they seem to. This is not a radical revision of common sense. Second, the semantic theories that have this result, in one form or other, are far from universally accepted. They are not even widely accepted. This being said, there are two reasons why the argument is important. First, if the conclusion was true, it would effect an ironic and unexpected resolution of the debate between pragmatists and their foes about correspondence. Second, there is an intriguing connection between this issue and one in the philosophy of biology, which concerns the concepts of fitness and adaptation.

Correspondence as a property of thought, and adaptedness as a property of biological structure and behavior, are both apparently used in explanations of success. It is in accordance with common sense to say that Jenny succeeded in getting around a city because she had true beliefs about how it was laid out. It is a similar type of explanation, in some ways, to say that a given trait succeeded under natural selection because it made its bearers well adapted to their cold environment. In both cases there is a relation between an organic system and the world – correspondence, adaptedness – that purports to explain the success of some entity in an environment. There has been extensive discussion of adaptedness, adaptation and fitness. Lewontin, for example, has argued that the idea of adaptedness is a holdover from a theological world view; it is a teleological concept of "fit" between organism and world. In particular, it encourages a passive conception of the organism, a view in which the organism simply reacts to the environment without altering or determining it (1977b, 1982).

This should sound familiar – Lewontin's criticism of adaptation echoes Dewey's attack on correspondence. In particular, Dewey and Lewontin associate the concepts of correspondence and adaptation (respectively) with what I have been calling an asymmetric externalist picture of organism/environment relations.

There is the possibility that correspondence can be fitted into a naturalistic world view, but only by robbing it of its explanatory teeth. It becomes a product of successful action, rather than a cause. At least, this is one turn the debate about correspondence can take.

The analogy with adaptedness is interesting because this *is* the turn the philosophy of biology has largely taken in that case. It is common now not to regard adaptedness as some lock-and-key relationship between trait and environment, which can explain success under selection, but rather to see success under selection as an ingredient in the concepts of adaptedness and adaptation. An adaptation, roughly, is a trait that owes its prevalence to the operation of natural selection. A trait is adapted, in a specific context, if it is liable to succeed under selection. The explanatory role of these concepts has been reduced, because the only way it has seemed possible to make them naturalistically acceptable has been to analyze them in terms of success under selection.

The analysis of adaptedness in terms of success under selection, and the analysis of correspondence in terms of a link to behavioral success, both serve to dethrone concepts that purported, pretheoretically, to describe some important relation between organic systems and their environments. They do this because they define the relations of correspondence and adaptedness in terms of a link to success. If correspondence goes the way of adaptedness on this score, then, I will argue, this would be a vindication of pragmatism after all.

6.4 Explaining representation

A central project in philosophy of mind over the last few decades has been the attempt to give a naturalistic analysis of intentionality – the attempt to say, in physical or at least nonsemantic terms, what it is for an internal state of a system like a brain to represent or be about some other object or state of affairs. The bulk of this work has been focused on explaining the content of belief-like states in particular, as opposed to desire-like states. The discussion in this chapter will also focus on belief-like states, states which make a claim about how the world is.

It is not possible to give an extensive survey of this work here (see Stich and Warfield, 1994), but I will make use of two broad distinctions. The first is a distinction between *modest* and *immodest* theories (Sterelny, 1990). The second distinction, which cross-cuts the first, is a distinction between *success-independent* and *success-linked* theories.[12]

Immodest theories attempt to give a fully general analysis of representation in naturalistic terms. Modest theories try to divide and conquer. Modest theories only try to give a naturalistic account of the most basic representational capacities. Then theories which are not themselves naturalistic, as they presuppose representation in some form, can be used to explain more complex types of representation.

Ruth Millikan's theory (1984, 1989b) is one of the most immodest, as it uses the same apparatus to explain why inner states of beetles can be about mates, and why the English word "mate" means mate. It is also fairly immodest to try to explain all mental representation (including human beliefs) at once, leaving public language and the like for another theory. Fairly immodest theories include those of Stalnaker (1984), Papineau (1987) and Fodor (1990), if I understand them right. The most modest view is to hold that only the explanation of the representational "rock-bottom" should be expected to be fully naturalistic. These rock-bottom abilities will be abilities that humans share with other animals. Modest views are, in this respect, foundationalist in their approach to representation. Sterelny opts for a modest view.

For example, causal theories of reference might be good at explaining the semantic properties of proper names in languages like English (Devitt and Sterelny, 1989). But these theories presuppose more basic representational capacities at several points. The phenomenon of "reference-borrowing" requires that people have thoughts about other individual people, so they can intend to use a term to refer to the same person that another does. Causal theories of reference, and perhaps other theories such as "conceptual role" theories, can be understood by the naturalist as part of a package, accompanied by a modest theory explaining basic capacities of representation.

The fact that many of these semantic theories, both modest and immodest, assign content to states of organisms other than humans raises again the issue of whether these theories really are, as I claim, rehabilitations of correspondence as a property of thought. Some might see a sharp distinction between the type of semantic naturalism seen in causal theories of linguistic reference, and the type of naturalism exemplified by attempts to say that a frog's inner states are about the flies at which it instinctively snaps. Indeed, some theories extend their generosity even beyond the amphibians which are now familiar features of the philosophical landscape. The relation in Millikan's theory which is the basic analog of the traditional relation of correspondence is also a property of adrenalin flows. When an adrenalin flow is produced, and there really is a dangerous condition present that requires fight or flight, this adrenalin flow corresponds to the world. But an adrenalin flow is not even a *representation*, surely? So how can it correspond to anything?

I will address the frog question first, and then the adrenalin question. With respect to simple representational states of animals such as frogs, which are assigned genuine content by many naturalistic theories, the bullet must definitely be bitten. A frog can have states that correspond to the world, in basically the same sense in which the sentence "snow is white" does.

I say "basically" the same sense, because it may be best to say that correspondence is equally a feature of frog states and English sentences, but *truth* is not. Truth may be a particular type of correspondence, associated with linguistic representation. This point applies as much to portraits and road signs as to frog neurons – portraits and road signs can represent the world as it is, but perhaps they are not thereby true. There is the possibility that inner states of frogs are sufficiently unlike linguistic entities for it to be better to say that frogs can have states that can correspond to the world, but which cannot strictly be true.

Adrenalin flows pose a more difficult problem, as while there is an intuitive sense in which they are "directed" upon the world, a fact accommodated by Millikan's theory, they are not normally considered representations. Adrenalin flows are, biologically, supposed to occur in certain environmental conditions, and they "misfire" when they occur in other conditions, but there is an apparent difference between this simple biological relation and genuine representation.

Millikan herself (in correspondence) opts to bite this bullet also – there are true adrenalin flows, or at least, adrenalin flows that correspond in the sense in which pictures and other nonlinguistic entities do. For advocates of her theory this is probably the best way. But there are also more moderate views possible here. A theory can restrict its domain to entities with the basic functional role of belief-like inner states (Stalnaker, 1984; Fodor, 1987). This will enable us to distinguish belief-like states in frogs from adrenalin flows.

The important point here is that if correspondence is understood as representing things as they are, then given the way many contemporary naturalistic theories are set up we must be prepared to find correspondence in unexpected places. Whether truth, as well as correspondence, is found in frogs or not is a less important issue; if truth is tied to language then correspondence is the more theoretically basic concept.

The second broad distinction we will use is a distinction between *success-independent* and *success-linked* theories of representation.

Success-independent theories include "picture" theories of meaning, and theories based on the idea of an inner model of the world which can bear a relation of isomorphism to reality. This family also includes pure *indicator* theories of representation. These are theories which analyze representation in terms of some relation of correlation or law-based connection between inner states and their objects. The theory of Dretske's *Knowledge and the Flow of Information* (1981), and Fodor's "asymmetric dependence" theory (1987, 1990) are examples.

The relevant common element of these theories is that for all of them a belief has its content independently of the consequences of any behavior directed by that belief, independently of the consequences of behavior directed by beliefs of that type in the past, and independently of the consequences of behavior produced by the mechanisms associated with that belief. Success and failure are not essential parts of the explanation of representation in any way. This is because these theories determine the content of the inner state either by looking to some intrinsic structural properties of the representation (as a resemblance or isomorphism view does) or by looking to causal and nomic connections "upstream" of the representational state, such as connections to the world which go by way of the properties of perception.

Consequently these theories are able to preserve the common sense view of truth as a fuel for success. Further, they may be able to help us understand why truth should have this property. Having an inner model that is isomorphic to the world is having a useful resource for coordinating action with the environment. Basing action on inner states that are products of the normal operation of perceptual mechanisms also seems a reasonable goal. These theories are not only compatible with common sense on this score, they also potentially vindicate it. They analyze semantic properties in a way that makes desiring truth comprehensible.

6.5 Success-linked theories

The second group of semantic theories comprises the *success-linked* theories. These can be split into three subcategories.

(i) *Atemporal* success-linked theories,
(ii) *Teleonomic* theories,
(iii) *Indicator/Success Hybrids*.

The relevant common feature of these theories is their making some link between an inner state and behavioral success constitutive, or partially constitutive, of the state's having a given truth condition. Consequently all these theories reduce the content of the claim that truth is a fuel for success. Some trivialize it almost entirely, others just reduce its content slightly.

We will begin with atemporal theories. This view was first clearly expressed as a theory of belief-content by Frank Ramsey (1927), who calls it a "pragmatist" view.[13] This approach has also been developed more recently by Papineau (though he does not hold it, 1987), Appiah (1986) and Whyte (1990).

Whyte, for example, proposed the following formula as an account of belief-content.[14]

(R)   A belief's truth condition is that which guarantees the fulfillment of any desire by the action which that belief and desire would combine to cause. (1990, p. 150)[15]

Leave aside the issue of explaining what it is to satisfy a desire. In this discussion I will simply take "success" for granted, within each theory, as unproblematic. The basic idea of the atemporal view is that a belief's truth condition is whatever state of affairs guarantees that actions based on the belief will be successful. Truth conditions are success-guaranteeing conditions.[16]

Taking R at face value we see the basic idea of success-linked views of meaning in its simplest form. What formula R does is buy an assignment of truth conditions to inner states at the expense of the trivialization of the idea that truth generates success. A belief would not *be* the belief that *p* if it did not lead to successful action when *p* obtains. There is nothing left of the idea of a semantic relation between a belief and the world that is conceptually independent of success, and that might be viewed as a useful relation in bringing about success.

That is how things would be if *R* was taken at face value as a theory of belief content. But there are complications to note within Whyte's own 1990 proposal, and the situation with other success-linked theories is more complicated still.

Firstly, *R* as quoted above looks extremely strong. Whether truth is conducive to success or not, how can it literally guarantee it? In fact there are some idealizations operating. Whyte, as I understand him, is dealing with the case in which all the agent's relevant beliefs are true. This may not ever happen, but if it did happen success would be guaranteed. Content is fixed by a relation between truth and success that would exist under certain ideal conditions. So in actual cases, where not all beliefs are true, there is no necessary link between success and truth.

Given this there is the possibility of a weak explanatory role for truth. We could say that a person succeeded because what happened in an actual case approximated the ideal case in which a link between truth and success is absolutely necessary. But this explanation should not be confused with an explanation in which truth is a relation between thought and the world that *causes* or *generates* success. Rather, this is an explanation in which an event is shown to be similar to an ideal case in which a connection is trivial and necessary. This may be a real explanation. But it is not the type of explanation we have been considering.

The *teleonomic* approach to explaining content takes the basic relationship between truth and practical success seen in atemporal views and places it in an explanatory context. "Teleonomic" has the same sense here that it had in Chapter 1 – it has to do with explanations of the existence of things in terms of their effects. Content is not determined by present relations between the belief's behavioral role and success, but by facts about prior episodes. These can be episodes in which a belief of that type was produced, or episodes that explain the existence of the mechanisms responsible for the production and role of the present belief. The prior episodes resulted in success because a particular state of the world obtained when the organism behaved a certain way, and these prior episodes are said to explain the present existence of that type of belief or that mode of belief production and use. Also, in the teleonomic view, the concept of "success" that was unanalyzed or understood in terms of the satisfaction of desires in the atemporal view, is replaced by a specific concept of success under a regime of selection. The regime of selection can involve natural selection across generations, or individual learning, or various combinations of these mechanisms.

David Papineau (1987) defends a straightforward teleonomic view:

> The biological function of a given belief type is to be present when a certain condition obtains: that then is the belief's truth condition. (1987, p. 64)

The function of a belief of a certain type is whatever that belief type has historically done that explains its survival in the cognitive system of the agent. For Papineau, *being present* (and hence producing a particular behavior) *under certain circumstances* is something a belief type can do that will explain why it is there.

Another teleonomic theory, and the most detailed success-based theory of any kind, is the view of Ruth Millikan (1984, 1989b). The central concept in Millikan's view for us is that of an "indicative intentional icon." Inner

180

belief-like states are one type of indicative intentional icon, and public language indicative sentences are another. An indicative intentional icon has a role in the interaction of two cooperating devices. One is the "producer" of the icon, the other is its "consumer." Consumers have biological functions, and are supposed to modify their activities in response to icons that they receive from the producer. Any icon is produced in order to modify the activities of the consumer, and in general this modification can either lead to the successful performance of the consumer's functions, or it can fail to. Putting it briefly, the truth-condition of an icon is the state of the world required for the icon to affect the activities of its consumers in a way which leads to their performing their own functions (whatever they are) in a historically explanatory way.

The object of selection here is the set of mechanisms underlying that mode of icon production and consumption – the mechanisms underlying the physical relations between producer and consumer, determining how the latter reacts to the products of the former. So in the case of a belief-like state, the (roughly) perceptual end of the information-processing system (and part of the inferential mechanism and memory) is the icon-producer, and the part of the system that produces behavior is the consumer.

To be more precise, the content of an inner icon is determined by an interpretation rule that relates the specific form of the inner icon to conditions in the world. This interpretation rule is determined as follows. The consumer mechanisms have been selected to react in a specific way to the form of icons produced, in accordance with what we might call "consumption rules" (not to be confused with interpretation rules). This way of consuming icons has been successful, *because* of a certain relation between the state of the world and the ways icons affect action. This set of consumption rules coordinates actions with the world in a way that was selectively successful. The icon is only supposed to exist in such-and-such a form if the world is such-and-such a way – the way that generates success when the icon is interpreted by the normal consumption rule. The interpretation rule is the principle that links specific icons to specific conditions in the world. These conditions are the icons' truth conditions.[17]

In the case of atemporal success-linked theories of representation it was easy to see how the idea that truth is a fuel for success is trivialized. The situation is more complex for teleonomic theories. The chief complication is due to the fact that the success used to determine meaning in these theories is past success, and we assume here that we are trying to explain a present episode – an episode occurring after the inner state has acquired a truth-condition. Teleonomic theories do not, strictly speaking, assert any relation between truth and present success. On Papineau's view an inner state C acquires from its history the function of being present when S obtains. In these historical episodes, C being present with S led to beneficial consequences. But if the environment has changed, or other aspects of the organism have changed, C's present contribution could be indifferent or disastrous. Similarly, in Millikan's view a certain pattern of consumption of icons has been established because it has led to success in the past, and the explanation for this success involves a certain relation between icons and the world. But it is entirely possible for this pattern of icon consumption to be disastrous in the present conditions.

The situation with these teleonomic theories is in fact similar to that seen with the atemporal theories, however. There is the possibility of a weak explanatory role being played by truth, but it is not possible to keep the idea of truth as a fuel or resource which generates success.

What exactly is truth, on a view like Papineau's? It involves two things: (i) an inner state must be tokened that is of a type which has, historically, affected the rest of the system in a way that led to success under selection when a certain state of the world obtained, and (ii) that condition in the world must now obtain. For an inner state to be true is for it to presently be in conditions in which it has been helpful, in a certain way, in the past. In one sense, the desire for truth is comprehensible on this view; it is a desire for inner states that have previously been useful in the conditions that obtain now. It is a desire for the "tried-and-true." Certain types of explanation are also possible. We can look at a present episode of truth and success and locate this episode as a member of a pattern stretching into the past. But truth is not a substantive relation between thought and the world that is *generating* success.

The same applies on Millikan's view. Expressing the point is more complicated in this case, and in the next section I will look at some details of her theory's handling of this issue. But the basic point can be made in a simplified way now. A belief is true for Millikan if: (i) it has a form that is supposed to be consumed by its consumers in a specific way, a way that will lead to their performing their functions in a historically explanatory way if a certain condition obtains, and (ii) that condition obtains. A truth condition is a condition that has a historical association with success. Being true is actually being in the conditions that have a historical association with success.

To be *previously associated with success* (for some unspecified reason) is not to be a fuel for success. There is no substantive relation between sign and world which generates success in this picture because the *only* thing that can be said about the relation between inside and outside that determines truth is that it was success-generating.

In the case of atemporal theories, a weak explanatory role for truth is derived from the possible link between actual cases and an ideal case of perfect information. In the case of teleonomic theories, it derives from the possible link between present cases and cases in the past. In the teleonomic case there is explanation in the sense of a subsumption under a historical generalization, but no more. In the atemporal case there is explanation in the sense of comparison to an ideal case or limit, but no more.

The third type of success-linked theory of meaning I will look at is the *indicator/success hybrid*. An example is the theory of Dretske's *Explaining Behavior* (1988).[18]

This theory makes use of both indication (law-based correlation) and function. Function is again understood in a basically teleonomic sense.[19] Dretske claims that an inner state C represents external condition S if and only if C was recruited as a cause of some bodily motion M because of a relation of indication between C and S. So C has to both antecedently have a relation of indication to S via the causal channels of perception, and C also has to have been recruited as a cause of a particular behavior because of this indication relation. Dretske regards individual learning as the only "recruitment" process sufficient to make C into a genuine belief, though recruitment by natural selection across generations is sufficient to make C into a lower-grade representation, with the same content.

The process of "recruitment" Dretske envisages must be driven by some form of feedback – a process of selection or reinforcement establishes a link between C and M by virtue of the consequences of producing M in certain conditions. So the view has a success-based or at least reinforcement-based component, as well as an indicator component.

In Dretske's paradigm cases, production of M at the tokening of C leads to the organism obtaining a reinforcing stimulus. The reinforcement leads to the link between C and M being established as part of the functional architecture of the system. Although C *indicates* S antecedently of this recruitment, it does not *represent* S antecedently. Its representing S is a consequence of the recruitment, because this is what gives C an indicator function. Only after recruitment does C have the capacity to *mis*represent the world. Indication itself is not a semantic relation. It is, however, the unique raw material for representation, so I will say it is "proto-semantic."

Only by virtue of the process of recruitment does C have the ability to correspond to the world or fail to correspond to it. Success associated with the production of M at the tokening of C is required for this recruitment, so success is now an ingredient in C's having the ability to correspond to the world. State C is a true belief if (i) C bears a relation of indication to a specific external state, (ii) there is a pattern of past success leading to the recruitment of the inner state as a cause of certain behavior as a consequence of the indication relation, and (iii) the external state obtains.

In this theory the status of common sense claims about truth and success is still more complicated. Dretske's view makes use of a historical association with success, like Papineau and Millikan, but it also makes use of an independently specifiable relation between inner and outer – the relation of indication. So there is a *proto*-semantic relation that has the potential to play some role in explaining successful dealing with the environment, and genuine representation is made up of this proto-semantic relation plus a certain past pattern of success.[20]

Dretske's theory, because of its residual appeal to indication, does have the potential to preserve more of the idea that truth is a fuel for success than the other theories discussed in this section.[21]

6.6 Millikan's maps

Many of the naturalists discussed in this chapter propose theories that rehabilitate correspondence while avoiding the term "correspondence" in their presentations. Millikan is an exception, and she uses this term in a way that is apparently incompatible with what I have said about the role of correspondence in teleonomic theories. So in this section I will argue against what Millikan says about her own use of the concept of correspondence.

I have argued for a weak conception of what a correspondence theory is. There is no need for picturing or isomorphism between sign and object, correspondence is just representing the world as it is. Millikan's theory certainly counts as a correspondence view in this sense. But Millikan sees her view as correspondence-based in a stronger sense; she sees it as resurrecting the idea that true representations "map" reality. In fact, a sign's having content consists in its being "supposed to map" the world (1984, Chapter 6).[22] Further, she sees mapping the world as something that can exist independently of a process of selection. After a process of selection, there come to be inner states that are "supposed to map" the world, just as a process of selection can bring it about that an organ is "supposed to pump" blood around. Pumping exists as a capacity of hearts independently of selection; selection turns the capacity into a function. Millikan treats mapping in the same way, as a capacity of inner states that can be promoted to being a function.

If this is the right way to look at Millikan's theory, then her view can make more sense of the common sense view of truth and success than I said it can. "Mapping" would play a similar role to that played by indication in hybrid theories like Dretske's. Inner states would antecedently map, and be promoted into representations in virtue of selection processes; they would be promoted from mere *de facto* mappers into things that are supposed to map, just as Dretske has states promoted from de facto indicators into states that are supposed to indicate. Both theories would make use of a proto-semantic relation – indication for Dretske, mapping for Millikan – and would assert that this relation is promoted into a genuinely semantic relation by selection processes. Representation would consist in being *supposed* to have a proto-semantic relation (indication, mapping) to something in the world.

Dretske's view and Millikan's are more different than this though. Mapping does not have the role in Millikan's theory that the comparison with Dretske suggests. The difference is that nearly *any* relation can constitute a "mapping" in Millikan's theory. Not just anything can be indication – the concept has a clear meaning that can be described in nomic or probabilistic terms. There are some problems surrounding it, but the basic sense is clear. There must be a real correlation between indicator and indicated. If the crowd only rushes up to the bar when it is last call, then the rush indicates last call. If they will rush up when it is last call, but also when it is the end of happy hour, or when a person of known generosity is buying drinks, the rush does not indicate last call. Millikan's concept of mapping, as it functions in her theory, is not constrained and definite in this way. The only thing different mapping relations have in common is the fact that they can play a certain role in explanations of the selection of modes of icon production and use.

There are *some* constraints on the relations that can figure in patterns of selection of this type. Consequently there is one substantive property of representations that Millikan's discussions of mapping do assert. This is a property of systematicity. For Millikan, all representations are necessarily structured, not into parts but into variant and invariant "aspects." Any representation is related to other possible representations by changes to the variant aspects. In the simplest cases, the time and place at which the representation occurs are variant aspects of it. Though we might say an inner state has the content "mate" or "predator," this is in fact shorthand for something like "predator in the vicinity now." What are usually termed "indexical" aspects of these simple contents are for Millikan no different in kind from the articulation of sentences into subject and predicate, and other more sophisticated compositional properties of representations. But this requirement of systematicity is, as far as I can see, the only substantive property of representations that her discussions of mapping assert.

Examples will make this clearer. Consider an inner state of a fish, which is statistically relevant to the presence of appropriate mates. The fish can be selected to respond to the presence of this state in a certain way, and the state can become an indicative intentional icon. The inner "consumers" of the state will have certain functions (to get the fish into contact with mates) and the inner state will have a normal condition for affecting the activities of its consumers in a way that helps them fulfill their functions in a historically explanatory way. That normal condition is the presence of an appropriate mate, so that is the content of the inner state. There is a "mapping" between the inner states of the fish and the world. The presence of the inner state at time t_1 and place X maps onto the presence of an appropriate mate at t_1 around place X. When the inner state occurs at time t_2 and place X, this maps onto the presence of a mate at time t_2 and place X. This is the relation between inner states and the world that was salient in the explanation of the selection of that mode of icon production and consumption.

Now suppose the case is different. When the inner state occurs it is not statistically relevant to there being a mate then and there. But it is statistically relevant to there being a mate at the reef next door a day later. Then the inner state might be consumed in a different way – the fish is selected to go to the reef next door on the following day. The content of the state is different. Its presence at place X and t_1 means "mate at reef next door to X one day from time t_1." This also counts as a mapping. *Any* relation between inner and outer that can become salient in the selection of icon-use is a mapping.

A chameleon changes color to match the surface it is on. As Millikan says, its skin color is supposed to map, by a given rule, the surface. Consider an anti-chameleon, which has been selected to change color to something very different from the color of the surface it is on. This also is a mapping, and the chameleon can have in it devices that are supposed to produce a skin color that "maps" the surface, by this other rule. *Any relation between chameleon and surface, provided it is repeatable in a way that can lead to selective success, is a mapping.*

So in my view, Millikan's relation of "mapping" is *post hoc*, as far as explanations of success are concerned. It is not a resource specifiable in advance, which can play an explanatory role. Millikan cannot say, as Dretske can say in the case of indication: "look for relations of mapping between inner

186

states and the world – these relations are the ones organisms will find to be useful resources." Consequently, I think that some of Millikan's use of the traditional language of correspondence is misleading. Her view is not, as earlier views were, an attempt to explain representation with the aid of an antecedently existing relation of picturing, isomorphism or mapping between signs and the world.

6.7 A stock-take

It will be apparent that the situation with respect to the relation between naturalistic semantic theories and common sense views of the pragmatic desirability of truth is very complicated. Before going on it may be helpful to do a stock-take.

The view of meaning that has the strongest tendency to trivialize the idea of truth as a fuel for success is an *immodest* version of an *atemporal* success-linked theory. Trivialization is produced by theories that try to give an analysis of all representations (or all mental representations) with belief-like or declarative contents in terms of an atemporal link to success. As we proceed to teleonomic views, and then to indicator/success hybrids, and finally to success-independent views such as pure indicator views, we get more and more back of the idea that correspondence is a substantive relation between inside and outside that may have a role in generating success. Also, as we apply success-linked theories to smaller and smaller parts of the total realm of representational capacities – as we get more modest – there is progressively less of a problem with the deflation of the common sense view.

Let us focus specifically on the issue of modesty. There is an important difference between theories that explain content *compositionally* and those that do not. A compositional approach to belief content views beliefs as composed of parts that have their own semantic properties, such as concepts. Belief contents are built up out of the contents of these basic units. A non-compositional view assigns a truth condition to an inner state "all at once," without building it up from the semantic properties of its parts. There may be good reasons for viewing compositional structure as a feature only of more sophisticated types of representations, and not a feature of frog inner states.[23] Supposing this is so, it might turn out that our theory of the most basic types of representation is a modest success-linked theory, but our account of the richer representational capacities of humans is something different, and which is compositional. Compositional views, if they work at all for belief contents, have an easier time with the link between truth and success than noncompositional views do. If we can establish that Jenny has the concept CLIFF and the concept AHEAD then there is a good chance that we can explain what it is for Jenny to put these concepts together in a way that amounts to her having the belief that there is a cliff ahead, without requiring that she necessarily behave in a way that is successful if there is a cliff ahead. That is, it may be that explanations of success in terms of correspondence only have their common sense status where they have their common sense home – in the domain of intelligent thought and action by normal human agents. It may well be that naturalism requires that we give up a set of intuitions about low-level cases of representation, in order to make sense of meaning as a whole.

Within the project of giving a modest theory of basic representational capacities, Millikan's teleonomic theory seems to me the best candidate proposed so far, although it is not without problems (Pietroski, 1992). Millikan presents her theory in an immodest way, but that does not prevent others from making use of it more modestly.[24]

This is not to say that if we had a satisfactory package of semantic theories, the common sense view of truth and success would automatically be vindicated. I have discussed the possibility that the need for a naturalistic theory of meaning actually prevents the vindication of common sense. But even once a theory of meaning which is not strongly success-linked has been put in place, it will be a further task to show exactly how, within this theory, truth manages to generate success.

6.8   A flurry over fitness

At this point we will leave, temporarily, the issues surrounding correspondence and look at some analogous questions in the philosophy of biology. I will argue that a similar set of problems have been encountered here, concerning the explanatory role of certain properties of organic systems in relation to their environments. There are the properties of adaptiveness and fitness. Like correspondence, these are relations that appear, prima facie, to have a role in explanations of success. However, when people have tried to say more clearly what these properties are, they have often drawn so heavily on the relation between fitness and success that success has become partially constitutive of fitness or adaptiveness, or at least an ingredient in it. So these properties have tended to lose their capacity to causally explain success. The literature in the philosophy of biology has moved further in one of the directions in which discussions of correspondence in the philosophy of mind could move.

The question of correspondence

The literature on fitness, adaptiveness and adaptation is vast, and I will make no attempt to survey it.[25] I will sketch the course of one part of the discussion though. I think it is agreed by many that in nineteenth and early twentieth century evolutionary thinking, the concepts of "fitness" and "adaptedness" meant roughly the same thing. They both denoted some relation between organisms (or parts of organisms) and their environments which exists independently of any property of evolutionary success, and which might in fact explain evolutionary success. Individual animals which are well adapted tend to stay alive, find mates and reproduce more effectively than others. Fitness in an organism is some property of good design or aptness to environment, and the fittest will survive.

The meaning of both concepts has shifted since then, perhaps for different reasons. First, "fitness" acquired a specific mathematical role in population genetics. It became a relative measure of reproductive output, associated in the first instance with genotypes of individual organisms (Roughgarden, 1979). Population genetics is often regarded as constituting the theoretical core of modern evolutionary theory, and in standard population genetics relativized measures of reproductive output are central to what the theory does. It does not much matter where differences in reproductive output come from; the theory describes what will happen to the population as a result.

The apparent identification of fitness with high reproductive output generated an ongoing discussion, prompted by the recurring allegation that Darwinian theory is explanatorily empty, as the phrase used to describe its basic mechanism, "the survival of the fittest," has no empirical content (see for example Popper, 1974, p. 137). Philosophers eventually succeeded in finding a way to answer the tautology charge while keeping intact the conceptual connection between fitness and reproductive output. This is the popular "propensity view of fitness" (Mills and Beatty, 1979). In the propensity view, the fitness of an organism is its *expected* reproductive output, where the expectation is calculated with probabilities understood as propensities (probabilistic dispositions). This is now the single most popular view of fitness, though it also has critics (Rosenberg, 1985; Byerly and Michod, 1991). On the propensity view, having a high fitness is not the same property as actually having a large number of offspring. But it is almost the same property: it is being disposed to have a large number of offspring.

Let us note what has become of the explanatory link between fitness and evolutionary success. The evolutionary success of some type of organism can be explained in terms of its having high fitness, but only in the weak sense in which the existence of a disposition explains its manifestation. Saying that

189

a trait did well because of its fitness is like saying a glass broke because of its fragility. The explanations are conceived, on the propensity view, as having the same structure and status. They are not completely empty. But saying that a trait succeeded because it had high fitness is not saying anything more than: what was liable to happen, actually happened (Sober, 1984).

Note again that the explanations of breaking in terms of fragility and of success in terms of fitness are different from the explanations of the breaking of the glass in terms of the glass' specific physical composition, the properties that make it fragile, and the explanation of the success of the organism in terms of the structural properties that make it likely to survive and reproduce, such as camouflage or metabolic efficiency. Fitness, like fragility, cannot be identified with any particular structural basis of this type, as fitness and fragility are multiply realizable. What makes a butterfly fragile is not what makes a champagne glass fragile, and what makes a whale fit is not what makes a nematode fit. If an identity is desired, fitness can be identified with the property of having *some* structural property or properties that make the organism likely to have a lot of offspring.

The situation with the concept of adaptiveness is not as clear, but I will try to give a similar rational reconstruction of what has happened (see also Burian, 1983, and entries in Keller and Lloyd, 1992). I said the concept was initially similar to that of fitness, involving a relation of well-suitedness between organism and environment. It was – and still is – used to describe parts of organisms, and specific behaviors, as well as whole organisms. A pattern of coloration or a territorial strategy can be said to be adaptive, and it can also be said that polar bears are well adapted to the Arctic.

One original role the concept had was that of an *explanandum*. For Darwin and others, adaptation was a visible feature of the organic world. Darwin thought that a satisfactory theory of evolution had to account for "the innumerable cases in which organisms of every kind are beautifully adapted to their habits of life" (Darwin, 1969, p. 119).[26] But in addition to this role as explanandum, individual differences in levels of adaptiveness became part of the Darwinian *explanans*. When some individuals are better adapted than others to local conditions, and the phenotypic properties that make this so are heritable, evolutionary change results.

Though the concept of fitness became mathematicized during the twentieth century, adaptiveness did not. It remained a less precise concept, purporting to describe some abstract biological relationship between organism and environment, a relationship often described with metaphors.[27] The concept was similar in some ways to the pretheoretic concept of correspondence as a property of thought. Both were abstract relations between the inner and outer, conducive to some kinds of success.

Like correspondence, adaptiveness and adaptation have been attacked as concepts which distort our understanding of organism/environment relations (Lewontin, 1977b). Whether because of the pressure of attacks like these, or because of a vaguer uneasiness about the idea, the concept of adaptiveness has subsequently been analysed by many in a way that quite transforms its theoretical role. The concept of adaptiveness has been brought back towards that of fitness – towards the *new* conception of fitness based on reproductive output. This is clear, for example, in Sober's 1984 account (Sober, 1984, p. 174). Sober takes adaptiveness to be interchangeable with fitness, and fitness is understood according to the propensity view. Some others do not go this far, but it is fairly common now to understand adaptiveness in terms of a dispositional property linked to selective success: a trait is adaptive in an environment if it is prone to be selected in such an environment. An animal or plant is well adapted to an environment if it is likely to survive and reproduce in it. So adaptedness has emerged with the same impoverished explanatory role that fitness has.

I have not said much yet about the relation between being adaptive and being an adaptation. There is reasonable consensus on what it is to be an adaptation; this is widely regarded as a historical concept. Sober's 1993 definition is a fair expression of the consensus view:

> Characteristic *c* is an adaptation for doing task *t* in a population if and only if members of the population now have *c* because, ancestrally, there was selection for having *c* and *c* conferred a fitness advantage because it performed task *t*. (1993, p. 84; see also Brandon, 1990)

The concept of adaptation has become analysed along the same lines as the teleonomic sense of function. An adaptation is something that has a Wright-style, teleonomic function. This function is *t* in Sober's definition. When this view is accepted it is natural to understand adaptiveness as an atemporal, dispositional analog. To be adaptive is to be disposed to be selected, and to be an adaptation is to have actually been selected for.

There are various ways to clean up the details, but whichever way it is done, according to the trend I am describing, the only substantive concept here is that of *being selected*. That is what lies behind adaptiveness, fitness and adaptation, although it is modally and temporally modified in different ways. All of these concepts, which appeared to describe relations between organism and world which exist antecedently to evolutionary success, and which might

explain evolutionary success, have come to be analysed in terms of evolutionary success.

As I said earlier, there is not a strong consensus about the concept of
adaptiveness. But some dissenting views are otherwise very much in accord
with my analysis of the situation. Byerly and Michod, for instance, do not
accept a selection-linked concept of adaptiveness, but their response is to
dispense with adaptiveness as a single, general property of organisms: "there is
no theoretically useful concept of overall adaptiveness" (1991, p. 3). They
accept that organisms have various "adaptive capacities," such as metabolic
efficiency and camouflage, and these capacities are part of what determines
fitness, in the sense of the rate of reproduction of a type of organism. But there
is no such thing, in their view, as a general relation of adaptiveness between
organism and environment. So adaptiveness is sometimes, when not understood directly in terms of selection, simply ditched.

6.9   Significance of the two trends

Spencer regarded the idea of correspondence between inner and outer as
a central part of an account of the relation between organic complexity and
environment. Dewey thought correspondence has no place in a naturalistic
picture. In early sections of this chapter I argued against Dewey. Correspondence need not be seen as an alternative to the practical goals of dealing with
or transforming the world, as it can be instrumental in achieving these goals.
The orthodox naturalistic view of correspondence, which sheds talk of picturing and focuses on the concept of accuracy as a real and useful property of
signs, promises to make sense of this claim. If this is right, it amounts to the
defeat of an important line of argument in the pragmatist tradition.

But for correspondence to have a real role in the production and explanation of success, it must be conceptually distinct from the fact of success.
Success-linked theories of meaning threaten this independence.

Success-linked theories of representation are strongly analogous to selection-based theories of fitness and adaptiveness in philosophy of biology. In
both cases, a relation which appeared, prima facie, to be an explainer of success
becomes linked constitutively to it. Success-linked theories of representation
are nothing like consensus views, and there are dissenters from selection-based
theories of fitness also. My own tentative bet is that the right theory (or
theories) of representation will not be highly deflationary of the common sense
view of truth, when applied to humans. As outlined earlier, I favor a modest
theory of basic representational capacities, perhaps close to Millikan's, which

can be combined with richer and probably compositional theories of the content of human thoughts. So on balance, I would suggest that Dewey was wrong here.

But let us suppose for a moment, hypothetically, that the right theory of representation is substantially deflationary of the idea of truth as a resource for success. What would that mean for the opposition between Spencer and Dewey, and more contemporary debates between orthodox forms of naturalism, and pragmatism?

The answer is that this would be a vindication of pragmatism. Pragmatists attacked a certain family of epistemic values as superstitious, or as implicit denials of the practical and glorifications of the lives of the leisured classes. I have examined a specific line of reply here: some traditional epistemic values have a more general-purpose role. They are not tied to any particular set of interests and can be used in a variety of projects. In particular, whatever it is that you want to do in the world, it is best to be guided by representations that represent the world as it really is. But if correspondence winds up being analysed in terms of success, this reply is largely empty. In that case what we have is a supplanting of the old goals with pragmatist ones. Correspondence is retained as a way of talking in an old-fashioned way about certain representations which are valuable on the newer pragmatist criteria.

Success-linked semantic theories provide a way for naturalists to back unexpectedly into something like the outlook on meaning and truth associated with people like Davidson (1984). These are views which see meaning and hence truth as properties that we ascribe within social life to serve certain interpersonal ends. These ascriptions are necessarily guided by a principle of "charity"; when interpreting an agent's beliefs and utterances it is inevitable that the interpreter see the agent as holding mostly beliefs that the interpreter regards as true, and as using these beliefs rationally. Although Davidson does regard his view of truth as a "correspondence" view in a sense, this is not a view that retains the idea that correspondence truth is a physical relation between signs and the world that agents can reasonably desire because of the practical success it tends to cause. Dennett's "instrumentalist" view of meaning is a more naturalistic relative of this one (Dennett, 1981).

If naturalists back into a view of this sort, the situation would also resemble that envisaged by some philosophers of science. Hacking (1983) thinks that Peirce, Lakatos and Putnam (Putnam *circa* 1981, at least) all describe a fallback position for people who believe that inquiry develops in some positive way but who reject correspondence theories of truth. "One takes the growth of knowledge as a given fact, and tries to characterize truth in terms of it" (1983,

p. 56). This is one way of viewing Peirce's (1878) definition of truth as what is accepted at the limit of inquiry. That move is structurally similar to what is done, in a more moderate way, by success-linked semantic theories. In particular, it gives up the idea that correspondence truth exists antecedently and can explain success, as other philosophers of science have thought (Boyd, 1983). There is no need to regard this as the "world well lost." But something big has been lost.

The idea here is not that we would be pushed towards giving up the idea that success can be explained, and that it can be explained in terms of properties of thought. What would be given up is the idea that there is an informative and general explanation of success in terms of a special relation of truth, correspondence or accuracy (Stich, 1990). If this is so, there may still be other general-purpose fuels for success, such as probabilistic coherence in the Bayesian sense (though this can be analyzed in a toothless way also). Alternately, it may be that there are no *general*-purpose fuels for success with respect to thought. We would be left with a variety of cognitive strategies pursued by agents, which work well or badly in specific circumstances, and whose working can be explained in terms of the specificities of the situation. We could find ourselves accepting a sort of "explanatory nominalism," or what Rorty (1982) calls "anti-essentialism." A diverse range of individual cases have their own specific details, and there is nothing informative that they have in common.

If both success-based semantics and selection-based views of adaptiveness were accepted, we would find a striking convergence which would suggest, though it would not establish, a general moral. It would then appear that we have acquired a habit, visible in several domains, of believing in general, abstract relations of fit, well-suitedness or concord between the internal structure of organic systems and the external world. These relations purport to be practically useful. But these are also concepts we have inherited from earlier world views. Correspondence as a cognitive virtue *was*, as Dewey says, originally part of a conception of mind and knowledge which disparaged practice and intervention in the world. Fitness and adaptation *were*, as Lewontin says, part of a theological conception of nature in which everything has its natural place and has no business venturing outside it. Now we have a different view of the place of organic systems like ourselves in the world. It is of course appropriate to seek explanations of success, and highly general explanations should be sought if possible. But in the case of fitness we seem to be left with, on the one hand, the facts of reproductive output and selection, and on the other hand, an indefinite multiplicity of individual explanations of the ways in which particular organisms get by, but no general *and* explanatory relation of "adaptiveness." In the case of correspondence we may find ourselves in the same boat, with the facts of practical success and scientific advance, and with a multiplicity of explanations for why particular belief patterns and cognitive strategies turned out to be effective, but no general and explanatory relation lying behind all or most of these cases of success.

If this turns out to be the outcome (and, again, this is hypothetical) there would be little reason to keep using the concept of correspondence. If it has no real role to play, and can be no more than an antiquated honorific with dubious habits of thought hanging off it, we would probably be better off without it.

6.10   Summary of Part I

I will now give a brief summary of a view of the environmental complexity thesis that emerges from the preceding chapters as a good one.

The environmental complexity thesis can be understood in teleonomic and instrumental ways. In teleonomic form it is an adaptationist hypothesis, and an externalist view. It would be rash to make a strong claim about the truth or falsity of this hypothesis here, but the view is a coherent and promising one. This part of the investigation will be continued in Part II. In particular, we will look again at one of the most promising aspects of Dewey's version of the environmental complexity thesis, the idea that thought has most value in environments characterized by a combination of variability in distal conditions that matter to the organism's well-being, and stability in the correlations between these distal conditions and more proximal and observable conditions.

In instrumental form the environmental complexity thesis is a claim about what cognition is good for and what role it plays in life. The instrumental version is too strong as it stands, as discussed in Chapter 4. However, the instrumental version of the thesis can reasonably be seen as capturing a large *part* of the instrumental role of cognition in the world.

Complexity in environments is understood as heterogeneity. There is no single measure of environmental complexity; any environment will be simple in some ways and complex in others. The internal properties of organisms play a role in making particular complexity properties relevant to them or irrelevant. Conceding this is not giving up on the environmental complexity thesis.

Externalist views are not necessarily asymmetric externalist views. Dewey's view, and some other pragmatist positions, can be understood as asserting an *interactionist empiricist* conception of thought, a view in which constructive relations between mind and the world are stressed, as well as externalist explanations of properties of thought.

The environmental complexity thesis is consistent with the view that thought functions in changing the world, but this constructive activity of thought in the world should be understood in a narrow way. The constructive activity of thought is a particular case of the general phenomenon of organic construction of environments. Thought constructs the world by making changes, via action, to the intrinsic properties of external things.

The issue of whether a relation of correspondence has a role to play in developing the environmental complexity thesis is undecided. This is a consequence of the fact that there is no widely accepted naturalistic theory of what correspondence is. Existing pragmatist arguments against correspondence are not effective in the light of contemporary views of meaning and truth – correspondence can be viewed as a general-purpose fuel for practical success. But this "orthodox naturalistic" view of the role of correspondence in practical life must be abandoned to the extent that the right theory of representation is success-linked. So a question-mark remains over the concept of correspondence.

The view of mind endorsed here combines elements both from the pragmatist tradition and from recent naturalism. A focus on environmental complexity is a point of contact between the two traditions. The role given to reliability properties is characteristic of the naturalism of Dretske, Goldman and others. Interactionism, on the other hand, is a central theme in Dewey's pragmatism. Those are the three central components of the view of mind endorsed here: properties of environmental *complexity*, which make mind worth having, *reliability* properties, which give thought its purchase on the world, and a naturalistic *interactionist* view of the causal traffic between mind and the rest of nature.

Notes

1 Contemporary naturalists most often trace their lineage back to Quine (1969) but not to Dewey. See Kitcher (1992) for a good account of the recent history and contemporary geography of naturalism.

2 There are some very difficult questions here concerning the status of "deflationary" views of truth (Field, 1986; Horwich, 1990). The idea that all that correspondence involves is saying that the world is some way, when it is that way, is apparently captured by the schema T: "p" is true if and only if p. On the deflationary view, truth is not so much a real property of signs, but a device used by us to make certain kinds

of assertions in an indirect way. These views respect the *T*-schema. Am I bound to say that any theory of truth that respects the *T*-schema is a correspondence view, even if it does no more than respect the *T*-schema? This is a difficult issue. My position is that any theory of truth that both (i) respects the *T*-schema, and (ii) is "robust" or "substantive," treating truth as a genuine property of signs, is a correspondence view.

What exactly is a "robust" view? Robust views understand the property *being true* as having the same ontological status as other everyday relational properties, such as *being the daughter of*, and *being larger than*. Horwich describes the idea behind these views as follows: "Just as the predicate, 'is magnetic' describes a feature of the world, *magnetism*, whose structure is revealed by quantum physics... so it seems that 'is true' attributes a complex property, *truth* – an ingredient of reality whose underlying essence will, it is hoped, one day be revealed by philosophical and scientific analysis" (1990, p. 2). Horwich is opposed to such views. I am uncertain if having a robust view requires that one believe that truth can actually be reduced to something else, or have its "essence" revealed by analysis. This is certainly sufficient, and I will assume this view throughout this chapter. I am not sure if this is strictly necessary – perhaps establishing some weaker link between truth and other natural properties would suffice to justify the robust view (see also Devitt, 1991a, 1991b).

On the view assumed here, certain constructivist antirealist positions count as correspondence views, which some will find objectionable. If this is undesirable, it is not hard to fix: we can say that a correspondence view is one which (i) respects the *T*-schema, (ii) is robust, and (iii) has a realist metaphysics.

3 I conjecture also that some thinkers who have sympathy for pragmatism might regard correspondence as a concept associated with simple "instructive" styles of externalist explanation, of the type seen in older forms of empiricism and Lamarckian evolutionary views. These are mechanisms by which the environment leaves its mark or imprint directly on the organic system. In contrast, selective views envisage the environment as posing problems and determining the success and failure of variants which are initially generated by other means. A good solution to an environmental problem need not be modeled on the environment's pattern. It might be argued that correspondence has its home in theories asserting a "direct" effect of the environment, such as Spencer's view. However, if correspondence is conceived in the way outlined in this chapter, this association between correspondence and instructive mechanisms is not a necessary one. Correspondence conceived as accuracy can equally well be the product of a selective mechanism. Some theories of correspondence we will look at *require* the operation of selection, in fact.

4 Life is "[t]he definite combination of heterogeneous changes, both simultaneous and successive, in correspondence with external coexistences and sequences" (Spencer, 1855, p. 368).

197

Why does Spencer move to the language of correspondence in place of the language of balance and adjustment? Spencer sometimes seems to think that the language of "adjustment" is problematic as it may be teleological. At one point he lists examples of living responses to external events, and contrasts them to the "responses" given by dead or inorganic objects. In the case of the nonliving,

> we do not perceive any connection between the changes undergone, and the preservation of the things that undergo them; or, to avoid any teleological implication – the changes have no apparent relations to future external events which are sure or likely to take place. In vital changes, however, such a relation is clearly visible. (1855, p. 267)

Rather than analyzing the idea of a self-preserving or compensating change directly, Spencer uses the prediction of future changes to mark off the living, and his concept of correspondence often has a predictive character.

5 Spencer thinks correspondence marks off the living from the inanimate, but he is aware of the problem posed by the fact that there are many nonliving systems in which internal changes are closely *causally* linked to external changes. What more is needed for correspondence?

This problem is discussed with the aid of an interesting example, the "storm-glass." This is a vessel containing a solution which will crystalize into some feathery shape at an external change in temperature, where the shape formed depends on the particular temperature (1855, p. 372; 1866, pp. 77–78). There is a definite correlation between inner and outer. But Spencer holds that this is not correspondence, and that it is really on par with the motion of a straw blown by the wind, rather than the changes seen in a simple form of life.

If he chose, Spencer could deal with this case with an appeal to the concepts of homeostatic response and feedback that he heads towards elsewhere. He could say that the storm-glass does not respond to temperature changes in such a way as to counteract the effects of such changes, or to preserve the existence and individuality of the storm-glass. Instead Spencer analyzes correspondence in terms of prediction. He introduces the idea of a "secondary" change in the organic system. (Spencer was forever making distinctions between "primary" and "secondary" changes in systems.) In a living system, when an environmental change E_1 occurs, and this is followed by internal change O_1, this change O_1 is followed by another internal change O_2. This change O_2 "anticipates" the *next* external change E_2, which will occur as a consequence of E_1. There is no correspondence between inner and outer unless there is a second internal change as well as the first.

Spencer's problem has only been deferred. Suppose there is a sequence E_1, O_1, O_2, E_2 of the type outlined. Why does the second internal change O_2 count as an "anticipation" of the second external change E_2? Why is it not just an effect of the first internal change, that happens to coincide in time with the external change E_2?

The question of correspondence

At this point Spencer's exposition becomes unclear. He refers to the secondary change O_2 as being related to E_2 with "a certain concord in time, place or intensity" (1855, p. 373). Later he says there is correspondence when the internal set of changes "has reference" to the external; that a relation between one set of changes "implies" a relation in the other. These formulas are not helpful.

6 This is one among several strands in James' arguments against correspondence or "copy" views. See James (1909), especially pp. 216–217. This argument has also been used recently by Stephen Stich (1990). This is a better argument than James' attempt to show that the common sense concept of truth can be *analyzed* in pragmatic terms.

7 Spencer, in fact, at one point reaches a view about truth that makes use of both correspondence and utility.

> And lastly, let it be noted that what we call *truth*, guiding us to successful action and the consequent maintenance of life, is simply the accurate correspondence of subjective to objective relations; while *error*, leading to failure and therefore towards death, is the absence of such accurate correspondence. (1872, p. 85)

8 See also Dewey (1941), and the discussion of the Dewey/Russell debate in Burke (1994).

9 It is hard to work out exactly what Dewey's view(s) might be here. In *Logic* Dewey does admit a role for statements of "facts" in inquiry. But their role in this process cannot be described in terms of simply representing the world as it is. The "validity" of supposedly descriptive parts of inquiry "depends upon the consequences which ensue from acting upon them" (1938, p. 166). Dewey discusses physical maps in the same terms (1938, p. 397).

The treatment of these issues in *Logic* might be extreme. In *Human Nature and Conduct* (1922), at one point Dewey expresses a view which appears close to the one advocated here.

> It is the first business of mind to be "realistic," to see things "as they are.".... .
> But knowledge of facts does not entail conformity or acquiescence. The contrary is the case. Perception of things as they are is but a stage in the process of making them different. (1922, p. 298)

Dewey would not understand this in terms of a correspondence between mind and external things, however.

We should note also that there are also older passages in Dewey in which a blunt James-style reduction of truth to utility and/or verifiability is asserted. See, for example, the discussion of truth in *Reconstruction in Philosophy* (1948, pp. 155–160, first published 1920).

For a more extended discussion of a "know your enemy" example in this connection, see Russell (1945, p. 824).

10 There is an interesting relation between the arguments of Stich (1990) and those given by G. E. Moore (1907) against William James' theory of truth. Moore urged a gap between truth and success to prevent James reducing truth to success (James' least plausible avenue of attack on correspondence views). Moore was attacking pragmatism. Stich wants to establish this gap in order to prevent people from valuing truth – his aim is to assert a (different) form of pragmatism.

11 See Field (1986) for a cautious defence of the view that deflationary theories are not able to accommodate the apparent role played by truth in the explanation of certain kinds of success. Putnam (1978) is another influential treatment of truth and success. See Devitt (1991b) for a reply to Horwich. Note also that the issue of whether truth can explain success is not the same as the issue of whether *only* truth can explain success, although it is closely linked to it.

12 I will not discuss the place of "conceptual role" theories of meaning here. Conceptual role semantics is a popular theory of meaning which takes into account the entire causal role played by an inner state, as it mediates between perceptual input and behavioral output (Block, 1986). Though many find conceptual role semantics a plausible approach to some types of meaning, it is not so often regarded as a good way to link an internal state to an external object of representation, or a truth condition. For this we need physical links to specific conditions in the world, and following one or the other chain of physical involvement, the perceptual or behavioral, has seemed more promising. Once we have a theory which links signs to objects, however, it may be that a conceptual role theory is part of the total package of theories explaining meaning.

13 This view of content is related to claims made earlier by Peirce (1877). See Wiener (1949) for more background on this "pragmatic" conception of belief, and a link to the views of Alexander Bain (1859).

    Ramsey has a success-linked view of belief content but also uses the "inner map" analogy discussed earlier. And he is regarded by some as an early "deflationist" about truth also (although see Field, 1986). I will not try to work out how his view as a whole relates to the concerns of this chapter.

14 Whyte does not still hold this 1990 view (personal correspondence). I use it here as a clear illustration of the atemporal success-linked approach.

15 This should be understood to say: *if* there is an action which the combination would cause, then *p* is the action's success-condition. Maybe there is no action which that particular combination could cause.

16 There is a problem of detail with the pure success-linked approach to meaning, which is irrelevant to the argument in the main text. What is done about world conditions implicated in the success of *every* action? Whyte's example is the existence of terrestrial gravity. The best treatment of this is Millikan's, and it can be exported to other views. She says that there are some conditions required for the success (or normal functioning) of every action or organic response, but these can be abstracted away from the truth condition just because they are common to all

success-conditions. The truth condition of an inner state is the world condition which is peculiar to the success-conditions associated with that inner state in particular.

17 Different success-linked theories make different specific claims about the relation between a truth condition and behavioral success, over and above the basic differences between the atemporal and teleonomic approaches. Whyte said the truth condition is that which *guarantees* the success of action based on the belief – it is a sufficient condition for success. Papineau I interpret to have the same view here, or perhaps the view of Ramsey who said the truth condition is the condition necessary and sufficient for success. Millikan on the other hand says the truth condition is that *required* for the icon to further the consumers' functions in an explanatory way. The truth condition is apparently a necessary condition. In fact there is less of a gap between Millikan's views and the others here. The truth condition is that responsible, in a causal sense, for the selection of that mode of icon production. It is not just a necessary condition. See Godfrey-Smith (1994b) for more detail here.

18 Some of Papineau's expressions are reminiscent of indicator/success hybrid theories, as they suggest that a correlation between sign and object is needed, as well as a link to success (1987, p. 89). I take these expressions to be inessential to Papineau's view. If not, his theory is closer to that of Dretske (1988).

   Stalnaker (1984) gives an indicator theory that also makes use of a link to action, but the link to action is not part of what fixes belief *content*. It determines that an inner state is a belief, rather than some other attitude or state (p. 19). So this is actually a pure indicator view.

19 Dretske professes agnosticism on the analysis of function. But see Godfrey-Smith (1992) and (1993) on this point.

20 I am indebted to Bill Ramsey here, who made this point about Dretske's claims about the role of meaning in explaining behavior. I have adapted his point to apply to explanations of success.

21 Earlier I outlined some of the details of Spencer's discussion of correspondence. The link between what Spencer was aiming at and some modern naturalistic ideas about representation is surprisingly close. Here is another of Spencer's examples of a paradigm living response:

> If a sound or a scent wafted to it on the breeze, prompts the stag to dart away from the deer-stalker; it is that there exists in his neighborhood, a relation between a certain sensible property and certain actions dangerous to the stag, while in its organism there exists an adapted relation between the impression that this sensible property produces, and the actions by which danger is escaped. (1855, p. 374)

Recall the discussion of "primary" and "secondary" changes in note 5 above. In the stag example there are two external states, the sound E_1 and the possible danger E_2. There are also two internal states, the perception of the sound O_1 and the motion of fleeing O_2. The sound E_1 will cause the perception O_1. But the perception also bears an "adapted" relation to O_2. It causes O_2, and as a consequence of this the danger E_2 is escaped.

The stag's action tends to preserve it against the destructive consequences of E_2 – I take it this is what Spencer means by an "adapted" relation. But this is not the only relation between internal and the external states in a case such as this.

Dretske (1988) says that if an inner state O_1 became recruited as a cause of motion O_2 because O_1 is a reliable indicator of danger E_2, then O_1 is a representation with the content "Danger!" Dretske makes use only of one external state and two internal states, the perception and the behavioral response, while Spencer's example has two external states and two internal. But suppose the relation between E_1 and O_1, the sound and the perception, is granted as background, as a physically basic connection having to do with the impact of sounds on the system. Then the case fits into Dretske's analysis of representation, and O_1 has the content "Danger." So Spencer is here close to a *modern* analysis of correspondence!

Dretske requires for representation a reliable correlation between the inner state and the external state which is represented. Millikan's view does not require high reliability. Spencer sees some of the issues here: "How can the act of secreting some defensive fluid, correspond with some external danger that may never occur?" (1855, p. 373). The important issue is not the possibility of occasional failures in a generally reliable response. The problem cases are those in which there is apparently the relevant sort of coordination between the release of the defensive fluid and some danger, even if the statistical correlation is weak. If the defensive fluid is cheap, an animal may tolerate a large number of "false alarms," but this is the same type of adaptive coordination between the inner and outer as that seen when the mechanism is more reliable.

The external state E_2 can play a special role in the explanation of the link between the internal states O_1 and O_2 even if the correlation between these states and danger is poor. We can express this in Millikan's language: the link between O_1 and O_2 can only perform its biological function in accordance with a historically normal explanation if E_2 obtains in the environment. It is the episodes in which the motion O_2 enabled the stag to avoid an actual danger that explain the selection of this mode of response in the stag.

So some of the raw material that Spencer builds into a theory of life is the same raw material that Dretske and Millikan build into theories of representation. Spencer, and other proponents of strong continuity, would approve.

22 Millikan is prepared to use still more contentious language: inner states and processes are "pictures" which can "mirror" the world (1984, pp. 233, 314). See also Millikan (1986).

23 As noted earlier, Millikan's view seeks to be compositional, in an unusual way, about all representation. Here I am assuming a more standard, building-block view of compositional structure.

24 If Millikan's or some other teleonomic theory is understood modestly, this does still leave us with some of the most notoriously counter-intuitive consequences of these views. A molecule-for-molecule replica of a human, which sprang into being spontaneously and had none of the normal historical properties, would not have thoughts with content. Many regard this as absurd. It would be indeed absurd to claim this being had no mental life at all, but it is not absurd to claim that it does not have thoughts which represent external things. There are certain first-person experiential properties, or "qualia," associated with being a believing agent – feelings of conviction or resolve associated with images and sentences, and so on. The replica has these qualia along with all the others (pains, sensations, etc.). But one can have these qualia without having states with real representational content.

25 Some relevant works are Mills and Beatty (1979), Sober (1984), Rosenberg (1985), Brandon (1990), Byerly and Michod (1991), and various entries in Keller and Lloyd (1992).

26 Adaptation was a central explanandum also for the Natural Theology tradition, which influenced Darwin. According to this view, "there is an adaptation, an established and universal relation between the instincts, organization, and instruments of animals on the one hand, and the element in which they are to live, the position which they are to hold, and their means of obtaining food on the other..." (Charles Bell, quoted in Coleman, 1971, pp. 59–61).

27 Dobzhansky, for example, used musical metaphors. He spoke of adaptedness in terms of "harmony," and said that organisms are "attuned to the conditions of their existence" (1955, pp. 11, 12). Pianka is quoted by Byerly and Michod as seeing adaptiveness as "conformity" between organism and environment (Byerly and Michod, 1991, p. 6).

PART II

Models

7

Adaptive Plasticity

7.1 The question

An organism confronts an environment which has a range of alternative possible states. The organism itself has a range of possible states, a range of possible behavioral or developmental choices. The alternative environmental states have consequences for the organism's chances of surviving and reproducing, and the right organic choice for one environmental state is not the right choice for another. The organism receives imperfect information about the actual state of the environment, as a consequence of correlations between environmental conditions which matter to it and environmental conditions which directly affect the periphery of its body. Under what conditions is it best for the organism to make use of this information, and adopt a flexible behavioral or developmental strategy, choosing its state in accordance with what it perceives, and under what conditions is it best for the organism to ignore the information, and always choose the same option, come what may?

This chapter and the next will discuss this problem with the aid of some simple mathematical tools. The aim of the present chapter is to describe abstractly some of the circumstances in which it is best to be a smart, flexible organism and some circumstances under which it is best to be unresponsive and rigid. This should tell us something about the value of cognition, as cognition is conceived here as a device making possible extensive flexibility and adaptibility to local conditions.[1]

Cognition is one type of device making individual flexibility possible, but it is not the only kind. Organisms which alter their developmental trajectories in accordance with environmental cues exhibit a similar type of flexibility. In both cases, an environmental cue affects the state of the organism, and this cue is used to determine which of a range of options the organism adopts. In both cases we can ask the question: under which environmental conditions is it best for the organism to be flexible? Dealing with problems via perception, information-processing and behavior can be understood as a particular instance of a more general phenomenon: dealing with environmental complexity via flexibility.

The framework used in this chapter and the next is adaptationist, driven solely by costs and benefits, so its application to evolutionary questions must be viewed with caution. Some of the philosophical issues which occupied us in earlier chapters will be abstracted away from. Others will be embedded in the mathematics and examined in detail. In particular, we will look again at the role of cognition in environments which exhibit a balance between the variable and the stable.

7.2 Biological background to the basic model

Models of the value of individual flexibility have been discussed in detail in recent years within biological investigation of "phenotypic plasticity." I will make use of some basic ideas from this field. Phenotypic plasticity is the capacity of a single genotype (genetic type of individual organism) to produce a variety of phenotypes – a variety of structural, physiological or behavioral forms. One reason why this field is useful for us here is the prevalance of c-externalist patterns of explanation in both theoretical and experimental work in this area: a standard explanation given for the existence and/or maintenance of phenotypic plasticity is the value of this plasticity in a variable environment.[2]

Examples from recent literature illustrate this. Drew Harvell (1986) investigates defenses against predators produced by colonial marine invertebrate animals called "bryozoans," or sea moss. The bryozoans Harvell studies are able to detect the presence of predatory sea slugs, making use of a water-borne chemical cue. When sea slugs are around the bryozoans grow spines. The spines have been shown to effectively reduce predation, but also to incur a significant cost in terms of growth, so they are detrimental when sea slugs are not around. This example is especially appropriate for us as though it is far from the realm of high intelligence it does have a proto-cognitive character. There is here a rudimentary form of perception, sensitivity to an environmental condition which is not itself practically important but which gives information about a more distal and important state. The organism produces an adaptive response to the distal state.

Another example of an "inducible defense" is a response in fish reported by Brönmark and Miner (1992). Carp in Sweden can find themselves in a pond which contains predatory pike, or a pond which does not. When pike are around, the carp alter their patterns of growth and hence body proportions in a way that makes them the wrong shape to fit inside a pike's mouth. This altered body shape has a cost, as it creates additional drag while swimming. Lastly and more poignantly, Stearns (1989) discusses a snail which responds to the presence of dangerous parasites by altering its whole life plan. When parasites are around it switches from a long-term reproductive strategy to a short-term, "live fast, die young" strategy.

Many writers make a direct link between phenomena such as these and cognition. Usually the comparison is made with learning (Roughgarden, 1979, p. 217; Via, 1987, p. 50; Sober, 1994). That is, the phenotype of an organism is considered analogous to the entire behavioral *profile* of an organism. But in terms of the framework set up in Chapter 1, producing spines in response to predation is a *first-order* plasticity. It is a change to the organism's state, not a change to the rules or conditionals that govern the organism's changes of state. The sea moss would be doing something more akin to learning if it individually changed the circumstances under which it produces spines. That would be a second-order plasticity.

This distinction is not always easy to apply to particular cases. The reason I invoke it here is to stress that a model of adaptive plasticity need not be considered applicable to cognition only in the case of learning organisms. The basic question is: when is flexibility a valuable thing, and when is it not? When does heterogeneity in the world make heterogeneity in action or structure worth having, and when is it better to meet a heterogeneous world with a bluntly single-minded reply?

7.3 The basic model

The model discussed in this section has been presented independently by Nancy Moran (1992) and Elliot Sober (1994), and some other biological models have a similar structure (Cohen, 1967; Lively, 1986). Sober presents it as a model of the advantages of learning, while Moran applies it to plasticity in general. My presentation will differ from both Moran's and Sober's, in part because in the next chapter I will embed this model within another mathematical framework, signal detection theory, and my presentation is designed to make the links between the two frameworks clear. The model of plasticity is also closely related to a Bayesian model of experimentation (section 7.6 below), and very likely the core of the model can be found in models in various other decision-theoretic fields such as economics and operations research.[3]

We assume that there are two states the world can be in, S_1 and S_2. These states are encountered by the organism with probabilities P and $(1 - P)$ respectively.

For now, the environment can be conceived as being structured either in time or in space. The difference can be important but it will not matter yet. The probabilities of these environmental states are understood in some objective way. Perhaps the world is divided into a spatial patchwork of S_1 and S_2. Alternately the changes may occur in time.[4]

The organism has available two phenotypic states, or behaviors, C_1 and C_2. We will use the term "behavior," but the model applies to various other forms of organic flexibility, as outlined above. C_1 and C_2 can be any structural or functional features of the organism with which payoffs or fitnesses can be associated. The payoff for producing behavior C_i when the world is in state S_j is V_{ij}. So there is a payoff matrix:

Table 7.1: Payoffs

	S_1	S_2
C_1	V_{11}	V_{12}
C_2	V_{21}	V_{22}

We assume that $V_{11} > V_{21}$, and $V_{22} > V_{12}$.

At this point let us look briefly at the relation between what has been set out so far and some issues discussed in previous chapters. It has been said that the environment has two possible states. This is a complex or heterogeneous environment, compared to the case where S_1 or S_2 is a fixed feature of the world. Some might say that we only find out that the environment is complex when we are given the information in the payoff matrix, and know that the distinction between S_1 and S_2 does make a difference. On the view defended in this book, this environment is recognized as complex, with respect to S_1 and S_2, in virtue of its intrinsic properties. This pattern constitutes *relevant* complexity for the organism in virtue of the properties of the payoff matrix.

The two behaviors, C_1 and C_2, must be distinguished from the *strategies* available to the organism. The three possible strategies are:

All-1: always produce C_1,
All-2: always produce C_2,
Flex: produce C_1 or C_2 depending on the state of an environmental cue.

The point of the model is to describe the situations in which *Flex* is the best strategy.

The side-by-side comparison of the three strategies should not be taken to imply that the model only describes a situation in which the three strategies are all available to an individual agent. This is one way to interpret it, but not the only way. There could be populations of individuals which reproduce asexually and pass their strategy genetically to their offspring. More complicated genetic details could also be added. A model like this could also be embedded in a context where there is cultural transmission of a phenotype or a process of individual learning. The basic model in this chapter describes a part – the cost-benefit part – of a variety of more detailed models.

The cue used by the flexible strategy is the state of an environmental variable which provides information about whether the world is in S_1 or S_2. This information is imperfect; there is only a partial correlation between the state of the cue and the state of the world. This information about the flexible agent's reliability can be expressed in a matrix also. The entries in the matrix are the probabilities, *given* that the world is in some particular state, that a specific response will be made.

Table 7.2: Response likelihoods, $Pr(C_i|S_j)$

	S_1	S_2
C_1	a_1	$(1-a_2)$
C_2	$(1-a_1)$	a_2

If the organism adopts one of the inflexible strategies, *All*-1 or *All*-2, then there still exist probabilities of "right" and "wrong" behaviors (behaviors with high and low payoffs for that state of the world). If the organism adopts *All*-1 the chance of producing the right behavior is P, and the chance of a wrong behavior is $(1-P)$.

Here are the expected payoffs of the three strategies:

(1) $E(All\text{-}1) = PV_{11} + (1-P)V_{12}$

(2) $E(All\text{-}2) = PV_{21} + (1-P)V_{22}$

(3) $E(Flex) = P[a_1V_{11} + (1-a_1)V_{21}] + (1-P)[(1-a_2)V_{12} + a_2V_{22}]$

We should note an assumption made here that some may take to bias the case in favor of plasticity. We assume that plasticity comes for "free"; we are not imposing additional costs for setting up and maintaining the mechanisms that make plasticity possible. We are not charging the organism for any additional structural complexity needed to perceive the cue and produce a flexible response. This issue will be discussed in the next chapter. For now we simply note that this basic model makes the assumption of no extra costs for plasticity.

Before we ask about the conditions under which *Flex* is the best, we can determine first when one inflexible strategy is better than the other. *All-1* is better than *All-2* if and only if:

(4) $\quad P(V_{11} - V_{21}) > (1 - P)(V_{22} - V_{12})$

In signal detection theory the quantity $(V_{11} - V_{21})$ is known as the "importance" of S_1, and $(V_{22} - V_{12})$ is the importance of S_2. Sober introduces a useful term here. The *expected importance* of a state of the world is its importance multiplied by its probability. So according to formula (4), *All-1* is better than *All-2* if S_1 has a higher expected importance than S_2.

So if you are going to always do the same thing, you should do the thing suited to the state of the world with the higher expected importance.

Suppose formula (4) is true and S_1 has the higher expected importance. Under what circumstances will the flexible strategy be better than the best inflexible one? *Flex* is better if and only if:

(5) $\quad \dfrac{a_2}{(1 - a_1)} > \dfrac{P(V_{11} - V_{21})}{(1 - P)(V_{22} - V_{12})}$

Whether it is better to be flexible depends on whether the cue being used is reliable enough to overcome the difference between the expected importances of the two states of the world.

It is worth taking a moment to see how formula (5) works.[5] The terms on the left hand side are probabilities associated with the production of C_2. By assumption, this is the behavior suited to the state of the world with lower expected importance. For flexibility to be favored, the difference between the probabilities of right and wrong decisions to produce C_2 must outweigh the difference between the expected importances of the two states of the world.

So it is possible to think of the situation in the following way. The organism has the option of always acting as if the world is in the state with the higher expected importance. The model describes the circumstances under which the organism should ever be prepared to "change its mind," and produce the behavior suited to the less important state of the world.

Formula (5) can be used to establish basic constraints on situations where plasticity is favored. First we will discuss the special case where the cue used provides no information about the state of the world; the organism is only able to match its behavior with the world at chance. This is never better than being inflexible, and is almost always worse. To be matching behavior with the world at chance is for the probability of producing each C_i to be the same, no matter what the world is like. Then $a_2 = (1 - a_1)$. The left hand side of formula (5) is one. So the left can never be larger than the right, as we are assuming that the expected importance of S_1 is higher than that of S_2. The left hand side is equal to the right if the expected importances are exactly the same, and otherwise the left hand side is smaller and flexibility is positively worse.

The strategy of producing variable behavior but only matching it with the world at chance is sometimes called "bet-hedging" (Seger and Brockmann, 1987) or "coin-flipping plasticity" (Cooper and Kaplan, 1982). In psychology it is called "probability matching" when an organism which has no informative cue produces a variable response in which a_1, and hence $(1 - a_2)$, are set equal to P. So probability matching is one kind of bet-hedging. Some animals seem to do this in certain situations, and it has been hard, within adaptationist assumptions, to explain why they do (Staddon, 1983). For as formula (5) shows, this strategy is never better than inflexibility.

Probability matching can also be intuitive to some people. Suppose you are uncertain about whether to bother stopping at STOP signs when driving in a safe and familiar area. If there is a police car lurking in hiding it would be better to stop. But if there is not, it would be better to drive on through, if the way is clear. You have no information about whether there is a police car there or not on any particular occasion, though you have a good estimate of the overall frequency of there being a police car there. We suppose all you care about is the money involved in a fine and the annoyance, fuel and wear on the car from stopping without reason, and that one fine is equivalent to some large number of unnecessary stops to you.

Some people find it intuitive to think that in this situation it is OK to adopt a strategy of stopping some of the time, or most of the time, and being more likely to stop as your estimate of the probability of there being a police car goes up. Instead, the model shows you should always stop or never stop. It is not better to have a probability of stopping that "corresponds" to the probability of a salient state of the world.

As the previous sentence was intended to suggest, this result can be regarded as "anti-Spencerian." Probability matching is a way in which an organism can respond to a complex world by means of organic complexity, where a parameter describing the organic complexity "corresponds" to a parameter describing the environmental complexity; the organic parameter a_1 is set equal to environmental parameter P. This is a *possible* organic response to this situation, but it is not a *good* one. It is better in these circumstances for the organism to meet environmental complexity with organic simplicity and rigidity.

Later I will discuss a special case in which an organism has no information but does in fact do best to produce variable behaviors – a case in which equations (1), (2) and (3) do not apply. But let us move now to the case where the organism does have an informative cue.

If there is any correlation at all between a flexible organism's production of C_1 and C_2, and the states S_1 and S_2, we will say the organism is "tracking" the world.[6] Then $a_1 \neq (1 - a_2)$. We will use the term "bet-hedging" for the case when $a_1 = (1 - a_2)$. If the organism is tracking the world the cue is providing real information.

Even if the agent is able to track the world, formula (5) imposes fairly stringent conditions upon when tracking is better than not tracking. If one type of wrong decision is sufficiently disastrous, it may be best never to behave in a way which risks this error. In formula (5) this would be represented with a large asymmetry between the importances of S_1 and S_2.[7]

So when the world is complex and exerts a variety of demands upon an organic system, even when the organism need pay no cost to be flexible in response, and even when it has an informative cue to use in tracking the world, tracking the world is not always a good thing. Moderately reliable tracking of the world can be worse than not tracking it at all.

One way to think of this result is as establishing a "trajectory problem." It can be seen as akin to the much-discussed problem for evolutionary theory posed by intermediate stages of traits that are apparently only useful when fully developed. Even if eyes are highly useful in their present state, they have presumably had to evolve from no eye at all. So it has been asked: "what use is 5 percent of an eye?" One view is that 5 percent of an eye will probably be favored, if at all, for reasons other than its enabling useful sight. Others claim that even 1 percent of an eye is useful for seeing when compared to no eye at all, that 5 percent is better than 1 percent, and so on up (Dawkins, 1986). The model outlined here describes one type of case in which guiding behavior with 5 percent of an eye is worse than using no eye at all. In section 7.6, however, we will look at a similar model with the opposite moral.

7.4 The inducible defense case, part I

Earlier I mentioned an example of phenotypic plasticity investigated by Harvell (1986), in which bryozoan colonies grow spines in response to marauding sea slugs. This case, with various simplifications and modifications, will provide a basis for illustrations in this chapter and the next.[8]

There are two states of the world. S_1 will be the state of clear seas, and S_2 the state of sea-slug infestation. We will suppose that only one colony in ten encounters sea slugs. That is, $P = 0.9$ (I do not know if this is a realistic figure). The colony has two possible states, each suited to one state of the world. The choice is assumed to be an all-or-nothing one. The bryozoans either produce spines or they do not. Cases where the trait varies continuously, such as the *amount* of some resource invested in defense, are more complicated. C_1 is the normal form, and C_2 is the spined form; the colony's phenotype is what plays the role of a "behavior" in the model. Spines reduce the damage caused by sea slug assaults, but they also incur a cost in terms of growth. Here are some sample payoffs:

Table 7.3: Inducible defense payoffs

	S_1	S_2
C_1	10	0
C_2	8	6

The state of the world with the higher importance is S_2; the difference between deciding correctly and deciding wrongly is larger for S_2 than S_1. But since $P = 0.9$, the state with the higher *expected* importance is S_1. So if the bryozoan has to choose between one of the two inflexible strategies, it is better to choose C_1, and have no spines.

Under what conditions will a flexible strategy be favored? If the numbers given so far are substituted into formula (5), we get the following condition for the superiority of flexibility: $a_2/(1 - a_1) > 3$. Expressed informally, the requirement is that the bryozoan be three times more likely to decide to produce spines when this is the right thing to do, than it is to decide this when it is the wrong thing to do. So an acceptable cue could give the bryozoan a 0.9 chance of choosing to produce spines when this is right (a_2), but only if the chance of the cue being misleading and inducing spines in the absence of slugs ($1 - a_1$) is less than 0.3. Or the figures could be 0.75 and 0.25, and so on. If there is no cue this good available, the bryozoan does better to always produce the spineless, standard form.

At this point the reader might think that this is not very informative when we have no idea what types of cues are likely to be available, and where they come from. Clearly, in some circumstances a cue with $a_2 = 0.9$ and $a_1 = 0.7$ would be easily found and in others it would be a very tall order. Further, there are ways for organisms to shape the reliability properties of cues. These topics will be discussed in Chapter 8 below.

7.5 The precarious and stable, revisited

In Chapter 4 I used suggestions from Dewey to give a general characterization of environments in which cognition is useful. Cognition is useful in environments characterized by:

(i) *variability* with respect to distal conditions that make a difference to the organism's well-being, and by
(ii) *stability* with respect to relations between these distal conditions and proximal and observable conditions.

Without the unpredictable states in (i), cognition is not needed, and without the correlations in (ii) cognition cannot solve the problem. The model discussed in this chapter complements these ideas.

To ask about the conditions under which flexibility is favored is to ask what will make the left hand side of formula (5) large and the right hand side small. First let us look at the case where the importances (but not the expected importances) of the states of the world are the same. Flexibility is then favored when a_1 and a_2 are large, and P is intermediate in value, close to 0.5. This condition fits the Dewey-inspired requirement above. When P is near 0.5 there is unpredictability about distal conditions in the world. Flexibility requires also that there be good correlations between distal and proximal states. These correlations are described by a_1 and a_2. This is the predictability required by the second clause of the Dewey-inspired requirement. (Recall also the discussion of the Todd and Miller model in section 4.8 above).

We also need to look at the importances of S_1 and S_2. Flexibility is favored if the *expected* importances are similar to each other, and that depends on both the importances and on P. Flexibility is favored if any asymmetries in the importances are balanced or counteracted by asymmetries in the probabilities. If both are asymmetrical in the same way – if S_1 is both more important and also more probable than S_2 – then inflexibility is more likely to be better.

So it is not exactly right to say the model vindicates the claim that unpredictability itself, in distal states, is what is needed for cognition to be favored. That is true if both states are similar in importance. It is more precise to say that cognition is favored if there is (i) unpredictability in distal conditions, where this measurement is weighted by the importances of these conditions, and (ii) predictability in the links between proximal and distal.

216

7.6 Comparison to a Bayesian model of experimentation

In this section I will compare this biological model of plasticity to a Bayesian model of experimentation (Good, 1967; Skyrms, 1990). There are two reasons for doing this. One is the intrinsic interest of the links between different models of behavior and information-gathering. The other is the fact that the Bayesian model is usually presented in the form of a general justification for engaging in experimentation. The Bayesian model has the result that acting on information gained from an experiment can often make your expected payoff higher and can never make it lower. The biological model of plasticity describes some situations where using information from a cue *can* make the agent worse off. So it is necessary to clarify the relation between the models. There is no incompatibility; the difference stems from a different specification of the properties of a flexible agent.

Although the model of this chapter and the Bayesian model are both in a broad sense decision-theoretic, there are differences in the frameworks assumed. The Bayesian model is usually associated with the view that probabilities are interpreted as degrees of belief. The theory is aimed at describing the policies of a rational or internally coherent agent. The model of plasticity in this chapter is part of a theory of objective relations between states of organic systems and conditions in the world. Probabilities are understood in some physical way, not as degrees of belief. But the two models have, at bottom, the same structure.

The Bayesian model considers an agent facing a practical decision, who chooses acts that maximize expected utility. If each state of the world is an S_j and the payoff from performing action i in that state is V_{ij}, then the agent chooses the action i which maximizes $\Sigma_j Pr(S_j)V_{ij}$. The expected utility of the optimal act is also the expected utility of the decision problem.

Suppose the agent also has the option to engage in an experiment whose possible outcomes are taken to be informationally relevant to the states of the world involved in the decision. We will discuss a simple version of the model in which there are two states of the world S_1 and S_2, two possible actions C_1 and C_2, and the experiment has two possible outcomes o_1 and o_2. The agent's store of probability assignments includes the probabilities of o_1 and o_2 conditional on the two states of the world. When the agent observes the actual outcome of the experiment, the probabilities of S_1 and S_2 are updated in the light of this evidence, and the choice of action is reassessed. If some action is now superior the agent switches to that one. Otherwise they remain with their previously chosen action.

Models

The difference between the two models is that in the Bayesian model the flexible agent, the experimenter, can change his or her preferred action after receiving the cue but does not have to change it. The Bayesian agent observes the cue and then recalculates the best behavior in the light of it. The new best behavior might be the same as the old one. In the biological model of plasticity a flexible agent is bound to determine their action with the cue.

There are two senses in which the Bayesian agent cannot do worse and will often benefit from performing the experiment (as long as the experiment itself is free from expense). One sense is obvious. Consider the agent's expected payoff after the outcome of the experiment has been observed. Suppose the outcome of the experiment is o_1. The agent now uses a new set of probabilities for the states of the world, $Pr'(S_j)$, to work out the optimal behavior. These are identical to the conditional probabilities $Pr(S_j|o_1)$. The agent chooses the behavior C_i which maximizes $\Sigma_j Pr(S_j|o_1)V_{ij}$. Either this is a different act from the previous best one or it is the same. Either way, when the new and old actions are compared in the light of the new probabilities, the agent cannot be worse off. No action can have a higher expected payoff in the light of the new probabilities than the new optimum.

It is possible for the new expected payoff, after the experiment has been observed, to be less than the payoff which had been expected in the light of the *old* probabilities. Once o_1 has been observed the old probabilities are not relevant though. And further, the agent's expected payoff from acting on the basis of experimentation, before the *actual* outcome of the experiment is observed, is higher than or equal to the expected payoff in the absence of the experiment. This is the other sense in which a Bayesian agent "cannot do worse" from experimentation, the less obvious sense, described in a theorem proved by Good (1967).

The expected payoff from experimentation is the weighted average of the different expected payoffs resulting from optimal acts in the light of each possible experimental outcome. The weights are the probabilities of different experimental outcomes.[9] This expected payoff is always greater than or equal to the expected payoff from making a decision without experimentation.

The two models converge on a basic point. An experiment has no value in the Bayesian model if its possible outcomes make no difference to the rational choice of behavior – the same behavior is best whichever way the experiment comes out. These experiments also fail the test for being a good "cue" in the sense described by the biological model of plasticity. A cue that is not worth using in the biological model constitutes a worthless experiment in the Bayesian sense.

We will use the inducible defense problem of section 7.3 as an example. The probabilities are now viewed as degrees of belief, so the interpretation of the model is quite different. The agent's prior probability for S_1 is 0.9. With $Pr(S_1)$ at 0.9 and the matrix in Table 7.1, the best action is C_1, which has an expected payoff of 9. Consider an experiment with possible outcomes o_1 and o_2, where $Pr(o_1|S_1) = Pr(o_2|S_2) = 0.7$. This experiment does not have the capacity to change which action is the best one. If the outcome is o_1 then C_1 is the best behavior, with an expected payoff of 9.55. If the outcome is o_2 then C_1 is still the best behavior, with an expected payoff of 7.91. So if o_2 is the actual outcome, the expected payoff under the old probabilities is higher than the expected payoff under the new probabilities. In that sense o_2 is bad news. If we consider the point before o_2 is observed, however, the expected payoff with experimentation is the same as the expected payoff in the absence of experimentation, which is 9. The experiment has no value in the Bayesian sense, and it also fails the test for the usefulness of cues presented in section 7.3. If $Pr(o_1|S_1) = Pr(o_2|S_2) = 0.7$, then $a_2/(1 - a_1) = 2.33$, which is not high enough for the cue to be worth using.

Now consider a different experiment. It has possible outcomes o_3 and o_4, where $Pr(o_3|S_1) = Pr(o_4|S_2) = 0.9$. If o_4 is observed, C_2 is the better action. The expected payoff with experimentation is 9.36, where the expected payoff without experimentation is 9. The "value of the experiment" is the difference between these two values, 0.36.

An experiment with the capacity to change which behavior is optimal has a positive value. An experiment which cannot change the optimal behavior has a value of zero. No experiment can have a negative value in the Bayesian model. In the biological model of plasticity there is a sense in which an "experiment" can have a negative value. Agents in that model are constrained to either alter their actions in accordance with a cue or not. If the cue does not have the capacity to alter the rational choice of behavior, then using it to govern behavior reduces the expected payoff. The Bayesian model is a model of rational choice as applied to experimentation, while the model of plasticity is a model of the ways in which an organic system should have variation in its behavior governed by variation in cues deriving from the environment.

The Bayesian model makes use of a more "intellectualized" conception of the agent than the biological model does. Either framework may be the appropriate one, depending on the circumstances. Continuing the example from Dawkins that was mentioned in section 7.3, we find that 5 percent of an eye is always worth having if it comes for free and if its deliverances can be factored into choices in the way described by the Bayesian model. But

5 percent of an eye can easily be worse than no eye at all if the eye's role in the system is of the type described in the biological model of plasticity.

7.7 Another model using regularity and change

Within behavioral ecology there has been a recent focus on the evolution of learning. I will outline an example of this work which illustrates some of the themes this chapter has been concerned with.

David Stephens (1991) has investigated formally some relations between environmental regularity and environmental change, as they affect the evolution of learning. He notes that many writers have made a general claim that learning is an adaptation to environmental change (see, for example, Plotkin and Odling-Smee, 1979). This idea is motivated by what Stephens calls the "absolute fixity argument." If the environment never changes, and there are any costs associated with learning, then the best strategy is to genetically fix a single appropriate pattern of behavior. But Stephens also notes that other writers, such as Staddon, have claimed that environmental regularity is needed for learning to be favored. As the "absolute unpredictability argument" asserts, learning can only be useful if present experience is a real guide to the future. Otherwise there is nothing to learn. So what is the real relation between change and regularity in the evolution of learning?

Dewey said that thought is useful in a domain characterized by both change and regularity. I proposed a sharper version of this claim that distinguishes between variability in salient ecological conditions, and regularity in the relations between proximal and distal. Stephens investigates another type of relation between change and regularity.

Stephens distinguishes between environmental change *within* generations and change *between* generations. The environment's overall predictability is a composite of these two types of patterns, and each may play a different role with respect to the evolution of learning. Here is an intuitive suggestion: learning is most favored when *between*-generation predictability is *low*, so a single genetically fixed pattern of response is not useful, but *within*-generation predictability is *high*, so an organism's early experience is a good guide to what it will encounter later in life. Stephens presents a mathematical model designed to test this idea (see also Bergman and Feldman, in press).

In the model, two parameters represent the environment's predictability within and between generations. There are two resources in the environment, a variable one and a stable one. There are three possible strategies. Two are inflexible; one sticks to the variable resource and one sticks to the stable one.

The flexible or learning strategy samples the variable resource at first but only stays with it if it is in a good state. I will not discuss the details of the model but will summarise Stephens' results.[10]

Stephens' basic result is that the within-generation predictability is much more important than the between-generation predictability. For learning to evolve there needs to be some change, either between or within generations (as the "absolute fixity argument" said). But given that there is some change, the most important factor in determining whether learning evolves is the within-generation predictability. Learning is favored when within-generation predictability is high.

As Stephens concludes, it is probably a mistake to try to relate the advantages of learning to some single, overall measure of environmental predictability. It is inaccurate to say, for instance, that a moderately unpredictable environment, one neither wholly chaotic not wholly stable, favors learning. This is inaccurate because a moderately unpredictable environment might be one in which there is unpredictability between generations and predictability within, and where learning is hence favored; but it might equally be one in which all the unpredictability is within generations, and where learning is consequently useless. It is necessary to represent the structure of environmental change in more detail, and consider interactions between different *types* of variability and stability.

7.8 Extensions of the basic model: geometric means

The remainder of this chapter discusses extensions of the basic model of plasticity. The material in these sections is more technical and is not necessary for understanding the next chapter, although some of it is used in Chapter 9.

The Moran/Sober model of plasticity compares flexible and inflexible responses to environmental complexity by comparing the expected payoffs of these strategies, using formulas (1)–(3) above. Expected payoffs or utilities can play a variety of roles in decision-theoretic and biological models. They can be used to characterize "coherence" or rationality in a subjectivist model. They can also describe what will tend to happen if a decision policy is used a large number of times, or to predict the long-term prospects of a strategy in a process of selection or competition. But (1)–(3) can only be used in these predictive tasks if other conditions are met.

Suppose there is a long sequence of trials in which a strategy is used and various outcomes are experienced. Sometimes the environment is in S_1 and sometimes it is in S_2. If the actual frequencies of S_1 and S_2 in the sequence are close to the underlying probabilities P and $(1 - P)$, we will call this a "representative sequence" of trials. If the sequence of environmental states is representative, and some further conditions are met, formulas (1)–(3) can be used to predict the long-term payoff associated with a strategy.

The further condition we will focus on has to do with the relations between successive trials. Formulas (1)–(3) are predictive if the payoffs can be represented with constant numbers which are added to each other across trials. If there is a series of trials, the "balance" of fitness or utility will go up and down, but the amount added from a given type of right or wrong decision does not go up and down depending on the present state of the balance.

Not all payoffs and days of reckoning in life are like this. The interest paid on a bank account is calculated as a fraction of your balance, for instance. Here the adjustment made after a trial is done by multiplying your existing balance by some amount, or equivalently, by adding or subtracting an amount which is a constant fraction of the balance. Biological reproduction has the same multiplicative character, a fact which is important in several models of plasticity.

When the outcomes of trials are combined in a multiplicative way, the predictor of long-term success in a representative sequence of trials is not the arithmetic mean payoff, as in formulas (1)–(3), but the *geometric* mean payoff. To find the geometric mean payoff after n events, we multiply the outcome of each, and take the nth root of this product.

An actual finite sequence of environmental states need not be "representative," of course. There can be long runs of good or bad luck. When we give a geometric (or arithmetic) mean fitness using probabilities P and $(1 - P)$, we are only describing what will happen in a representative sequence. A very long sequence is very likely to be representative. But a more complete description is obtained by examining all the possible sequences of environmental states, including the unlikely ones.[11] In this book mean fitness measures will be used to predict long-term outcomes in an idealized way, by describing what happens in representative sequences.

If the geometric mean rather than the arithmetic mean is the relevant measure, this can affect the results of models of the type we have been discussing. An important difference between the two types of means is the fact that geometric means are more sensitive, in comparison to arithmetic means, to variance or spread in the numbers averaged. A few low values in the numbers tallied drag the geometric value down greatly. A single value of zero makes the geometric mean zero. We will look at several situations in which geometric means are the relevant measure.

Suppose, first, that an organism is making a sequence of decisions, and the entries in the payoff matrix are numbers that are multiplied, rather than added, across trials. Here are formulas for the geometric mean payoffs associated with the three strategies:

(8) $\quad G_1(All\text{-}1) = V_{11}^P V_{12}^{(1-P)}$

(9) $\quad G_1(All\text{-}2) = V_{21}^P V_{22}^{(1-P)}$

(10) $\quad G_1(Flex) = V_{11}^{Pa_1} V_{21}^{P(1-a_1)} V_{12}^{(1-P)(1-a_2)} V_{22}^{(1-P)a_2}$

Assuming all the payoffs are positive numbers, *All-1* is better than *All-2* if:

(11) $\quad P(\log V_{11} - \log V_{21}) > (1-P)(\log V_{22} - \log V_{12})$

If (11) is satisfied, *Flex* is better than the best inflexible strategy if:

(12) $\quad \dfrac{a_2}{(1-a_1)} > \dfrac{P(\log V_{11} - \log V_{21})}{(1-P)(\log V_{22} - \log V_{12})}$

The criteria for the superiority of flexibility are the same as they were in the model discussed earlier, except that logs of payoffs are substituted for the original payoffs. Bet-hedging is never favored over the best inflexible strategy, as before.

Some situations in which tracking of the environment is favored, under arithmetic and geometric optimization, are represented graphically in Figure 7.1. In these figures it is assumed that $a_1 = a_2$, for simplicity. The graphs represent the critical value of a_1 and a_2 against various values of P, for a given payoff matrix.

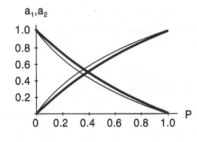

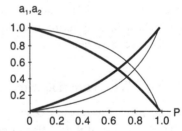

(a) Nearly symmetrical matrix

	S_1	S_2
C_1	3	1
C_2	1	2

(b) S_2 much more important

	S_1	S_2
C_1	3	1
C_2	1	10

Figure 7.1 a,b

The lines represent the lowest values of a_1 (and a_2) such that flexibility is better than an inflexible strategy. The two lighter lines represent the thresholds for the superiority of *Flex* over the two inflexible strategies, when arithmetic means are compared. *Flex* is only favored in the region above *both* lighter lines. Heavier pairs of lines represent the two thresholds when geometric means are used.

It should be kept in mind that the payoff matrixes play different roles in the two types of cases. Figure 7.1a, for example, compares a case where producing C_1 in S_1 results in 3 being added to the agent's balance of fitness, to a case where this event results in this fitness being multiplied by 3. The graphs show that in some situations higher reliability of tracking is demanded by arithmetic optimization than by geometric, and in other situations the opposite is true.

In some models of plasticity the use of geometric mean payoffs is associated with variation in time as opposed to space. But this is not a completely general relationship. If a model is describing what it is good for an individual, self-interested organism to do in the face of a variable environment, then there is no relevant difference between temporal and spatial variation. An organism might wander through different regions of space, or stay still and experience different events through time. Either way, at any moment it faces S_1 or S_2. Under both spatial and temporal variation, *either* arithmetic or geometric optimization can be relevant in predicting the long-term outcome. It depends on the structure of the situation, and what is "staked" by the organism at any one time; it depends on whether the present outcome is related to previous ones by multiplying or by adding.

You can walk through a casino, or stay put, and this does not affect the relevant predictor of long-term success. That is determined by the relation between your successive bets.

7.9 Variation within and between trials

So far the flexible organism has been one which varies its behavior between trials. It behaves differently on different occasions. Within any particular trial the organism does one thing only. What happens if an organic system can vary its behavior or phenotype *within* a trial?

Suppose the organic system can produce some proportion of C_1, and of C_2, on a single occasion. An organism might produce two different forms of a chemical simultaneously, each adapted to some particular condition it might be about to encounter. This has to be understood as a mixture of two discrete alternatives, not as a single behavior intermediate between C_1 and C_2. Note also that the world is not mixing S_1 and S_2 within trials; it is producing one or the other.

We will first look at the case where the same mixture is produced on every occasion. The organism is not really "flexible" but its behavior is heterogeneous or variable within a trial. As there is no correlation between behavior and the world, this is a within-trial analog of bet-hedging.

When there is production of a single mixture on every occasion and relations between trials are additive, the formula for long-term payoff is the same as that for the standard arithmetic case with bet-hedging, as discussed earlier in this chapter. That is, the formula used is (3), where a_1, which is equal to $(1 - a_2)$, is the proportion of C_1 produced. Mixing is never favored.

We will look now at a case which will be important in Chapter 9: the organism produces the same mixture in every trial, and trials are related to each other multiplicatively. Then the formula for the long-term payoff for variable behavior has some additive parts and some multiplicative parts. Let m be the proportion of C_1 produced within each trial.

$$(13) \quad G_2 = (mV_{11} + (1 - m)V_{21})^P (mV_{12} + (1 - m)V_{22})^{(1 - P)}$$

This case can be investigated by finding the optimal value for m given a specified payoff matrix and value for P. When m is zero or one, the organism is not mixing. Otherwise it is mixing. The optimal value, m^*, is given by the following equation (Cooper and Kaplan, 1982).

$$(14) \quad m^* = \frac{(1 - P)V_{21}}{(V_{21} - V_{11})} + \frac{PV_{22}}{(V_{22} - V_{12})}$$

Although there is no tracking of the world, mixing behaviors is often favored. Under the payoff matrix used in the bryozoan case, for example (Table 7.1), mixing of behavior is favored whenever P is above 0.8 and below 1. When $P = 0.9$, as in the example discussed earlier, the optimal value of m is 0.5.

Variable behavior is favored in this type of case because geometric means are reduced by variance in payoffs across trials. Varying *behavior within* a trial can be a way to reduce variance in *payoff between* trials (Seger and Brockmann, 1987).

It may also be possible to have variation both within and across trials. Different mixtures are produced on different occasions. Let m_1 be the proportion of C_1 produced in mixture 1, and m_2 be the proportion of C_2 produced in mixture 2. As before, a_1 and a_2 represent the degree of tracking the state of the world; a_1 is the probability of producing mixture 1 when the world is in S_1, and

225

a_2 is the probability of producing mixture 2 in S_2. If the relation between trials is additive, the general formula is as follows.

(15) $\quad Ad = Pa_1[m_1 V_{11} + (1 - m_1)V_{21}]$
$\qquad + P(1 - a_1)[(1 - m_2)V_{11} + m_2 V_{21}]$
$\qquad + (1 - P)a_2[(1 - m_2)V_{12} + m_2 V_{22}]$
$\qquad + (1 - P)(1 - a_2)[m_1 V_{12} + (1 - m_1)V_{22}]$

If trials are related multiplicatively, the formula is as follows:

(16) $\quad G_3 = (m_1 V_{11} + (1 - m_1)V_{21})^{(a_1 P)} \times ((1 - m_2)V_{11} + m_2 V_{21})^{(P - Pa_1)}$
$\qquad \times ((1 - m_2)V_{12} + m_2 V_{22})^{(a_2 - a_2 P)} \times (m_1 V_{12} + (1 - m_1)V_{22})^{(1 - a_2)(1 - P)}$

These doubly variable strategies can be compared both to wholly fixed strategies (where m_1 is one and m_2 is zero, or vice versa) and to single-mixture strategies (neither is zero but $m_1 = 1 - m_2$).

We will look at (15) first. In every case the best m_1 and best m_2 are zero or one; intermediate values of m_1 or m_2 are never optimal. In some cases both m_1 and m_2 are best set at one (these are the cases where the cue is an informative one) and in other cases either m_1 or m_2 is best set at zero.[12] So in this case the organism should either produce variable behavior across different trials without mixing within them, or (when the cue is not a good one) it should be completely inflexible.

Formula (16) is more complicated. In this chapter we are taking a_1 and a_2 as fixed, and the best pair of mixtures m_1 and m_2 is sought within this constraint. This is a problem of optimizing a function of two variables. In Chapter 8 we will also look at trade-offs between a_1 and a_2. If this trade-off is also under the control of the organism, the problem is more complicated still.

When a_1 and a_2 are fixed and the best m_1 and m_2 are sought, some degree of mixing within trials is favored under a wide range of conditions.[13] There are cases, however, in which it is best to produce no mixing within trials, and some cases in which it is best not to produce variable behavior at all, within or between trials.

Rather than discussing the general case in detail here, we will look at two special cases. First suppose the organism cannot track the world across trials: $a_1 = (1 - a_2)$. Then it can be shown that for any combination of mixtures there is a superior option which is producing only one of the two mixtures in question. This argument is parallel to the one used in (11)–(12) above, replacing "V_{11}" in (12) with the formula for the payoff for producing mixture 1 in S_1, and so on. So a single mixture is always better than two. Producing a single mixture is often better than not mixing at all, however.

Second, suppose that a_1 and a_2 are equal to one. Then (16) simplifies to:

(17) $\quad G_4 = (m_1 V_{11} + (1 - m_1) V_{21})^{(P)}((1 - m_2) V_{12} + m_2 V_{22})^{(1-P)}$

This is the same as a formula given by Moran (1992, formula 17) for the long-term fitness of a *genotype* which is expressed differently in different individuals within a generation, where the environment varies its state across generations. In this way a genotype produces mixtures within trials. Note that what must be envisaged is that within an S_1 generation there is one specific mixture produced, and within an S_2 generation there is a different mixture. There is tracking of the world across trials as well as a mixing within them. Proportion m_1 of the individuals with the genotype choose C_1 when the world is in S_1, and proportion m_2 choose C_2 when the world is in S_2. The fitness of the genotype depends on this action by individuals. Consequently, m_1 and m_2 are no longer decision variables which can be set at any level, but are constraints. Fitness is maximized by having m_1 and m_2 as high as possible, but in Moran's scenario this requires accurate tracking by individuals.

In formula (16), in contrast, it is envisaged that it is possible for the agent (organism or genotype) to produce the wrong mixture for the present trial, as well as the right mixture. Mixtures are not associated infallibly with states of the world, and a_1 and a_2 are constraints within which the agent must optimize. If a cue used by a whole generation, in the scenario Moran envisages, has the ability to mislead and lead to the wrong mixture in that generation, then (16) must be used rather than (17). And if the individuals do not track at all, so the mixture is the same in each generation, then formula (13) must be used (as in Cooper and Kaplan, 1982).

When (17) is used, high values of m_1 and m_2 always increase fitness. But let us take m_1 and m_2 as fixed, and ask when a given variable strategy will be favored over completely inflexible strategies, as represented in formulas (8) and (9). Figure 7.2 represents a simplified case, where $V_{11} = V_{22}$; $V_{12} = V_{21}$; and $m_1 = m_2$. Pairs of lighter lines represent the original thresholds for the superiority of flexibility under arithmetic optimization (and also, in this instance, for the superiority of flexibility in the "pure" geometric case of formulas (8)–(11)). As before, flexibility is favored in the region above both lines. Heavier lines represent the two thresholds in the scenario associated with formula (17).

Reliability of tracking can be very low while variability is still favored. In the high-importance case with $P = 0.5$, the critical value of m_1 and m_2 is under 0.25. Variability is favored even though behavior is strongly anti-correlated

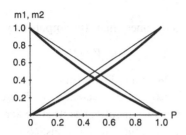 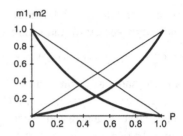

(a) Symmetrical smaller importances   (b) Symmetrical large importances

	S_1	S_2
C_1	2	1
C_2	1	2

	S_1	S_2
C_1	10	1
C_2	1	10

Figure 7.2 a, b

with the world. A higher correlation between behavior and world could be attained with an inflexible behavior. But because of the peculiarities of the regime, variability is better.

To sum up, when tallying across trials is multiplicative and there can be variation within a trial, a "naive c-externalist" expectation is more often fulfilled. Complexity in the world calls forth complexity in the organism. This can occur even if the organism is emitting this complexity blindly.

It must be stressed that these results only obtain when there is variation of behavior within a trial. So this applies to strategies pursued by individual organisms only if the individual can do two things at once, or if for some reason there is arithmetic tallying on one time-scale and geometric tallying on another, larger, scale.

The model also applies to "strategies" pursued by genotypes, or by groups. For example, a genotype can produce different phenotypes in different individuals of that type, and a genotype for flexibility can benefit even though there are some individuals with the genotype that do badly. Looking at the high-importance graph in 7.2, under this regime variable behavior is favored over fixed behavior when $P = 0.8$, as long as m_1 and m_2 are over about 0.6. On any occasion the world is 80 percent likely to be in S_1, but 40 percent of the individuals with the genotype (or in the group) produce the wrong behavior. We will look at this in detail in Chapter 9.

Notes

1 I will use the terms "flexibility" and "plasticity" interchangeably in this chapter – "plasticity" is an established term, but "flexibility" has a more convenient opposite in "inflexibility."

2 This is now a large literature. Some useful reviews and theoretical discussions are Bradshaw (1965), Schlichting (1986), Sultan (1987), Via (1987), Stearns (1989), West-Eberhard (1989), and Gordon (1992).

Sometimes the term "phenotypic plasticity" is reserved for cases where there is continuous variation in phenotype, and cases of discrete, all-or-nothing variation are called "polyphenism" (Moran, 1992). This terminological distinction, which is only sometimes adhered to in the biological literature, will not be adhered to here.

Within genetics, plasticity is often discussed in terms of the "norm of reaction" of a genotype. A norm of reaction is a function which assigns a phenotype to each environment, for a given genotype.

3 I have not undertaken a survey of models of flexibility used in these different areas. Some links are apparent from the biological literature. David Stephens (1986) presents a model of the value of information in the guidance of behavior which is explicitly derived from an economic model.

4 If we have here a sequence of events in which the environment is in different states at different times, these events are independent trials where the probabilities of the different environmental states are fixed. A more realistic model might use "Markov chains" of environmental states, where the probability of encountering a given state at the next step is conditional upon the state encountered in the present (and perhaps in the past). This is how Stephens' model (section 7.7) is set up.

5 We assume that S_2 does not have an importance of exactly zero, as then the right hand side of the formula is undefined. It is possible for a state to have an importance of zero, and if so, formulas (1)–(3) are used directly.

In most of this discussion I will compare *Flex* only to the best inflexible strategy, and will assume *All*-1 is the best inflexible strategy. This is for ease of presentation. Here is a more general formula only slightly altered from Sober (1994), for conditions in which *Flex* is better than both *All*-1 and *All*-2:

(6) $\dfrac{a_2}{(1-a_1)} > \dfrac{P(V_{11}-V_{21})}{(1-P)(V_{22}-V_{12})} > \dfrac{(1-a_2)}{a_1}$

If both inequalities are false (but neither relationship is an equality), then the cue is an informative one, but it is being used in the wrong "direction". It would be a useful cue if the pairing of behaviors with states of the cue was switched. In this discussion we will assume that the cue is being used in the right way. See also section 8.4, however.

6 This is a common usage in biology. In epistemology the term was used in a closely related sense in Nozick (1981).

7 The requirement in expression (5) can be expressed as a condition on posterior

probabilities, the probabilities of states of the world given states of the organism, via Bayes' theorem.

$$(7) \quad \frac{Pr(S_2|C_2)}{Pr(S_1|C_1)} > \frac{(V_{11} - V_{21})}{(V_{22} - V_{12})}$$

Bayes' theorem can be expressed in various ways. A relevant version here is: $Pr(A|B) = Pr(B|A)Pr(A)/Pr(B)$.

8 My discussion is very idealized. Clark and Harvell (1991) give a proper theoretical treatment of inducible defences, using dynamic programming.

9 A formula for the value of the decision problem with experimentation is $\Sigma_k Pr(o_k)(Max_i \Sigma_j Pr(S_j|o_k)V_{ij})$, where "$Max_i$" is the value associated with the C_i which maximizes the expression. The value of the decision problem without experimentation is $Max_i \Sigma_j Pr(S_j)V_{ij}$. This can be shown to be less than or equal to the value of the decision problem with experimentation. If the same behavior i is optimal under any experimental outcome, then the values with and without experimentation are the same. See Skyrms (1990) for a presentation of the theorem and a discussion of its role in different types of decision theory.

10 Stephens uses both analytical techniques and simulations. The model uses haploid genetics. There are alleles for learning, for always choosing the stable resource, and for always choosing the variable one. The variable resource has a good state in which it is better than the stable resource, and a bad state in which it is worse. The within-generation and between-generation predictability are modeled as two first-order Markov processes; the probability of the resource changing state at the next step depends only on its present state. Each generation only has two periods, so the environment changes either once in that generation or not at all. One problem with the model, which Stephens does mention, is the fact that learning has the effect of reducing variance in fitness across trials. As the calculation of overall fitness is multiplicative (expected logs of fitnesses used), bet-hedging is useful (see section 7.9). So the advantage of learning is conflated with the advantage of bet-hedging. This artificially increases the advantage of the learning strategy. I think the model would also be more informative if the generations were longer relative to the time-scale of environmental change. See also Bergman and Feldman (in press) for detailed discussion of genetic factors in this type of model of learning.

11 Another possibility is to regard the "probabilities" in the model simply as actual frequencies. Then the issue does not arise. The issue arises when the model is taken to describe a system in which probabilities are physical parameters, and actual frequencies of events can either conform or fail to conform to the probabilities.

12 If we take the derivatives of Ad (formula (15)) with respect to m_1 and with respect to m_2, in both cases we have a constant, which is positive or negative according to the relations between the a-terms, P and the payoff matrix. Fitness is either always increasing as a function of m_1 or always decreasing, and the same is true of m_2. The possible values of m_1 and m_2 are bounded by zero and one, so in every case the best m_1

and best m_2 are zero or one. The criterion for when m_1 and also m_2 are best set to one is formula (6), which describes when *Flex* is better than both *All*-1 and *All*-2 in the basic model.

13 To take one example, if the parameters in the bryozoan case are used, and $a_1 = a_2 = 0.8$, then the best combination is: $m_1 = 0.86$, $m_2 = 1$.

8

The Signal Detection Model

8.1   The next question

The previous chapter looked at some relationships between flexible and inflexible ways of dealing with environments while assuming that the cue used in a flexible strategy is a "given," a fixed constraint within which the organism optimizes. The organism either makes use of this cue, accepting its reliability properties, or it does not. It was also assumed that there are no additional intrinsic costs associated with being flexible. The only costs discussed in Chapter 7 are costs stemming from making wrong decisions. In this chapter both of these assumptions will be dropped. We will look at ways in which organisms can shape the reliability properties of cues they use, and we will also look at one way to build in some costs associated with the mechanisms that make flexibility possible. The same modifications to the model of the previous chapter will address both of these issues at once. In the previous chapter we assumed that a cue with certain reliability properties was available, and asked: should the organism use this cue in the determination of behavior? In this chapter the question is: given the general nature of an organism's physical connections to the world, what is the best cue available for guiding its behavior with respect to a particular problem?

These questions will be addressed with the aid of signal detection theory, a psychologistic application of statistical decision theory.[1]

8.2   Signal detection and the inducible defense case

To make the presentation of these ideas as intuitive as possible, I will use a concrete example from the start. This will be the inducible defense found in bryozoans, introduced in the previous chapter (Harvell, 1986). Bryozoans are colonial marine animals, some of which can detect the presence of predatory sea slugs with the aid of a chemical cue transmitted through the water. When sea slugs are around the bryozoans grow spines, which are an effective, but costly, defense.

There are two states of the environment, S_1 and S_2, where S_2 is the state of there being sea slugs. The probability of S_1, or P, will be set at 0.9. There are two possible phenotypes, C_1 and C_2. C_2 is the spined form and C_1 is the standard, nonspined form. Here is the payoff matrix used earlier:

Table 7.3. Inducible defense payoffs

	S_1	S_2
C_1	10	0
C_2	8	6

We will assume that the bryozoans can perceive the concentration of a chemical in the surrounding water. This concentration will be represented with a variable X. This variable gives some information about the state of the world. Putting it roughly, higher values of X are associated with the presence of sea slugs, although it is possible for high values to occur without slugs and low values with slugs.

Two "likelihood functions" describe the relation between the concentration of the chemical and the alternative states of the world. These functions, $F(X|S_1)$ and $F(X|S_2)$, both contain information about how likely a particular value of X is, *given* a particular state of the world. They are assumed to be "normal" or bell-shaped curves, as in Figure 8.1.

The height of the function $F(X|S_1)$ at some value, such as $X = 4$, does not tell us how likely that particular precise concentration of X is, given S_1. As X is a continuous variable with an infinite number of values, the probability of getting some exact value, given S_1, is zero. Probabilities are measured as areas below the curves. The functions do tell us how likely X is to have a value within a certain small interval, given S_1.

The *likelihood ratio* function, $lr(X)$, is the ratio between the values of the functions. That is, $lr(X) = F(X|S_2)/F(X|S_1)$. With the functions in figure 8.1, higher values of X always have higher likelihood ratios than lower values of X; this provides a more precise way of saying that high values of X are associated with the presence of sea slugs.

233

Models

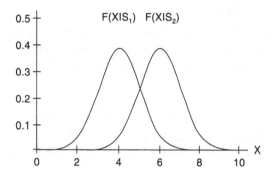

Figure 8.1. Likelihood functions.

Let us now consider the situation of a particular bryozoan colony which perceives a particular concentration of this chemical in the water. In the light of this observation, would it be better to produce spines or not? If we assume the same payoffs and value of P used earlier, we know that if the bryozoan is going to do one thing always, it should never produce spines. When is an observation good enough evidence of sea slugs (S_2) for it to be worth producing spines instead?

Signal detection theory finds a threshold value of X, such that if this value or a higher value is observed, then this observation is worth acting on.[2] By "worth acting on," I mean that the expected payoff from producing C_2 given this observation is higher than the expected payoff from producing C_1. In this chapter this will be only way in which strategies are evaluated. The observation received by the colony will be understood as perception of X falling within some small interval, rather than X having some precise value. What is observed is that X is within some small region (of size $2e$) around the value x.

Observations which are good enough evidence of sea slugs to motivate producing spines are observations for which the following is true:

(18) $\quad Pr(S_1|x-e<X<x+e)V_{21} + Pr(S_2|x-e<X<x+e)V_{22}$
$\quad\quad > Pr(S_1|x-e<X<x+e)V_{11} + Pr(S_2|x-e<X<x+e)V_{12}$

This condition rearranges to:

(19) $\quad \dfrac{Pr(x-e<X<x+e|S_2)}{Pr(x-e<X<x+e|S_1)} > \dfrac{P(V_{11}-V_{21})}{(1-P)(V_{22}-V_{12})}$

234

The fraction on the right hand side is the ratio between the *expected importances* of the two states of the world, which we encountered in Chapter 7. Formula (19) is similar to formula (5). In signal detection theory the right hand side of (19) is known as "β."

The final step is to get from formula (19) to a value of X that is set as the threshold. When e is very small, the ratio $Pr(x - e < X < x + e | S_2)/Pr(x - e < X < x + e | S_1)$ is approximated by the ratio $F(x | S_2)/F(x | S_1)$. So the likelihood ratio can be used to establish a threshold, such that all observations higher than the threshold satisfy the condition $Pr(x - e < X < x + e | S_2)/Pr(x - e < X < x + e | S_1) > \beta$. The best threshold is the value of X such that:

$$(20) \quad \frac{F(X | S_2)}{F(X | S_1)} > \frac{P(V_{11} - V_{21})}{(1 - P)(V_{22} - V_{12})}$$

The establishment of the threshold divides the possible observations into two classes, those which should be met with C_1 and those that should be met with C_2. The threshold establishes how good information has to be before it is good enough to warrant shifting from the default behavior to the alternative.

In the present example the value of the right hand side of formula (20) is equal to 3. So the threshold in Figure 8.1 should be at the point where the likelihood ratio is 3; where $F(X | S_2)$ is three times $F(X | S_1)$. This is about where $X = 5.55$. This point will be called X_c, the "critical" value of X. Once we have established the best possible setting of X_c, we have established the *best possible cue* for the organism, given the constraints that apply in the decision problem.

Once the threshold X_c is set, the probabilities of various types of good and bad decisions are determined. In particular, the values of parameters a_1 and a_2, which were important in the previous chapter, are determined. For a_1 is the probability of choosing C_1 given that the world is in S_1, or $Pr(C_1 | S_1)$. The organism should choose C_1 if the observed value of X is below the threshold X_c. So the proportion of the area under $F(X | S_1)$ that is to the left of X_c is equal to the probability $Pr(C_1 | S_1)$. That is, $Pr(X < X_c | S_1) = a_1$. In this case, with the threshold at $X = 5.55$, a_1 is about 0.94.

The value of a_2 is determined similarly. It is the proportion of the area under $F(X | S_2)$ to the right of the threshold. Here it is about 0.67. So if the threshold was moved further to the right, a_1 would increase and a_2 would decrease. More of the distribution for S_1 would be on the appropriate side of the threshold but more of the distribution for S_2 would be on the wrong side. Although the organism benefits from having both a_1 and a_2 as high as possible, it is not possible to maximize them both simultaneously. An improvement in one type of reliability must be purchased with a deterioration in the other.

So some reliability properties of a cue are subject to modification by the agent, but this modification takes place within constraints. An agent might be determined to avoid producing C_1 when the world is in S_2; this error might be absolutely dire. Then the organism can adopt a threshold far to the left hand side. That will make an erroneous production of C_1 almost impossible, but only by making the other kind of mistake, producing C_2 when the world is in S_1, very common.[3] Looking at formula (20), it is clear how the location of the threshold is dependent on the relationship between the expected importances of the states of the world. As the expected importance of S_1 goes up, so does the appropriate threshold.

We can illustrate this trade-off with another feature of the bryozoan case. Harvell found that smaller colonies were less likely to produce spines than larger colonies; small colonies "require a greater stimulus" (1986, p. 816). We might interpret this as the smaller colonies demanding better evidence of an impending attack (this is not the only possible explanation, but it provides a useful illustration). Suppose P is also unchanged. Then a different threshold could be explained in terms of differences in the payoff matrix.

Harvell's suggested explanation for the phenomenon is that the cost of producing spines might be greater for small colonies, if only because the spined periphery of a small colony is a larger proportion of the entire colony. Suppose this is so. Then we might alter the payoff matrix by adding to the cost of C_2:

Table 8.1: Inducible defense payoffs — small colonies

	S_1	S_2
C_1	10	0
C_2	6	4

If these payoffs apply, the two states of the world have equal importances. Once P is factored in, the best threshold is where the likelihood ratio is 9, not 3 as before. The threshold is shifted to the right, to where X is just over 6. Smaller colonies should demand better evidence of attack before they produce spines, as Harvell observed.

So the specificities of the problem determine the best way to make a trade-off between two opposing types of reliability. Reliability in one direction must be paid for with unreliability in the other. The only way to improve both types of reliability simultaneously would be to reduce the overlap between the two likelihood functions in Figure 8.1. For now we will continue to regard the overlap between the two as a fixed constraint. Later we will relax this constraint.

8.3 Optimal cues and acceptable cues

In Chapter 7 a definition was given of an *acceptable* cue, for an organism faced with an uncertain world. In the present chapter a definition has been given of the *optimal* cue for an organism. We have assumed that the organism gets a physical signal with certain informational properties, and a criterion has been given for the best way to use this signal to establish a decision criterion. We will look now at relations between these two models.

At this point it is necessary to start using the term "cue" in a more precise way, and to distinguish it from the physical signal the organism receives. The properties of the *signal* are given in Figure 8.1, while the properties of the *cue* are given when the threshold is specified as well. A cue is a signal as categorized by the organic system.

So, we have criteria for an acceptable cue, formula (5), and for the optimal cue, formula (20). In both cases there is a default behavior, C_1. The criterion describe circumstances in which it is worth making use of C_2. The right hand sides of (5) and (20) are the same – the ratio of the expected importances of the states of the world. The left hand sides of the formulas are both ratios associated with the probability of producing C_2. In the case of the model of plasticity, the criterion is given in terms of the overall reliability properties of the cue: $a_2/(1 - a_1)$. Signal detection theory gives a criterion in terms of a property of a threshold: $F(X|S_2)/F(X|S_1)$.

We would expect that the optimal cue should necessarily be one of the acceptable cues, but not vice versa. This is correct. If all the possible observations in the interval defined as the optimal cue have likelihood ratios at least as big as the ratio at the threshold, then all of these observations will, individually, be acceptable cues. The "weakest links" in the optimal cue are observations right near the threshold, and these weakest links are strong enough. So an optimal cue is always acceptable. In the bryozoan example, the ratio $a_2/(1 - a_1)$ for the optimal cue is just over 11, where 3 is all that is needed.

On the other hand, if a cue fails the test for optimality, it can still pass the test imposed by (5); it can still be better than no cue at all. To describe which suboptimal cues are still worth using, we need to distinguish between two ways in which a cue can be suboptimal. Recall that there is a "default" behavior; the behavior suited to the state of the world with the higher expected importance.

237

Here it is C_1. The other behavior will be called the "alternative." An overly *conservative* cue is one which leads to the production of the default behavior more often than the optimal cue would. If a cue is overly conservative then it must always be acceptable according to formula (5). All the values of X in the interval associated with this conservative cue have better likelihood ratios than the ratio at the optimal threshold.

An overly *adventurous* cue is one that leads to the production of the alternative strategy more often than the optimal cue would. An overly adventurous cue includes observations which, considered individually, are not sufficiently good evidence to motivate the alternative behavior. In some cases the cue as a whole will still be worth using despite this, and in some cases not; an overly adventurous cue may or may not be acceptable. In the example we have been following, if the threshold is set at $X = 4.49$ or above, the cue is overly adventurous but still better than nothing. If the threshold is less than $X = 4.49$ then a flexible strategy using this cue is worse than inflexibility. Though overly adventurous cues are not guaranteed to be acceptable, a specific overly adventurous cue can be superior to a specific overly conservative cue.

Signal detection theory can always define a cue which is worth using if it exists, but it defines this cue in terms of a likelihood ratio. There need not be anything in the world which the organism can perceive that is associated with this likelihood ratio. Sometimes not only the best cues, but also all the imperfect but acceptable ones, will not be possible cues in the situation, for the organism in question. Some reliability properties of the cue are at the disposal of the agent, but others are not; sometimes the world does not contain good signals of the environmental states that matter. And sometimes a good signal may exist but the organism will not be equipped to perceive it, or to categorize it in a way which produces an acceptable cue.

8.4 The costs of plasticity

At this point we must attend to an issue which was raised and then deferred in the preceding chapter. This is the idea that models of the advantages of flexibility must take account of the fact that, usually if not always, flexibility is *expensive*, compared to inflexibility. Aside from the costs associated with making wrong decisions, the mechanisms which implement a flexible approach to dealing with the world are often more structurally complex than the mechanisms which implement a brute, inflexible, dumb approach. One form of this is the idea that functional complexity requires structural complexity; that it requires additional machinery for perceiving cues and altering the organism's response.

This is, indeed, a common assumption (Sultan, 1992). Models often explicitly incorporate an intrinsic cost for plasticity, which must be balanced against its benefits. In some cases these assumptions have been very strong. In this section we will look firstly at the issue of costs when plasticity is considered in a very abstract way. Is there a general principle about cost which applies to plasticity in *all* its forms – plasticity in development, behavior, and so on? I will look at some counterexamples to claims that could been made along these lines. There are also more specific questions about expenses associated with various types of plasticity and the mechanisms implementing them. Whether or not plasticity is expensive in the abstract, cognition and certain other types of plasticity could be intrinsically expensive.

My attitude towards principles asserting *general* costs to plasticity is skeptical. The issue is an empirical one, and it would be inappropriate to make strong claims about it here. But as Sultan (1992) has argued, it is possible that several natural-seeming assumptions made about costs in this area are not so much obvious truths as questionable extrapolations from some features of information, specialization and cost within social and economic life.

Though I am skeptical about broad, general principles about costs, there is no denying that tracking the world and behaving flexibly is very expensive in many specific cases. So in the next section I will modify the signal detection model to incorporate one type of cost peculiar to flexible strategies.

Some assumptions which authors have made about costs of plasticity can be seen as applications of even more abstract principles about adaptation. A principle which has both significant intuitive force, and which has influenced thinking in this area, is the "Jack of all Trades" principle.[4] This principle asserts that a specialist on its own ground is always better than a generalist is on this ground. As the proverb says, the jack of all trades is a master of none, and will hence be outcompeted by a master when on the master's preferred turf. A flexible strategy is a generalist in the relevant sense, so it should do worse, by this principle, than some inflexible strategy in any particular environmental condition, though it might do better overall.

The Jack of all Trades principle would assert that a flexible strategy when dealing with environment S_1 must do worse in that specific environment than the strategy *All*-1. This can be understood in various ways. The flexible strategy might be unable to produce exactly the right phenotype in S_1. Alternately, perhaps the flexible strategy gets exactly the right phenotype, but pays an extra cost to do this. This is the way models of the advantages

239

of flexibility usually proceed, and a version will be discussed in the next section.

Although there is no denying its intuitive appeal, the Jack of all Trades principle is not something which has become established in the literature through explicit testing. Some writers have recently argued that the evidence for the principle is quite weak (Huey and Hertz, 1984; Futuyma and Moreno, 1988; Sultan, 1992). At least, when the idea is formulated in a way which is readily testable, the evidence for it is weak. Sultan argues that this biological assumption of the virtue of specialization may be linked to beliefs about the advantages of specialization and a division of labor in economic life – beliefs which became influential as a consequence of Adam Smith's *Wealth of Nations* (1776) and which were applied in strong forms in the automobile factories of Detroit.

An alternative way to make a case for a very general cost-of-plasticity principle would be to make use of the relation between functional and structural complexity. I said in Chapter 1 that we are concerned, in general, with functional complexity. The flexibility discussed here is a type of functional complexity; it is heterogeneity in what the organism does. It might be claimed that complexity in what is done requires structurally complex mechanisms underneath. If it was also claimed that complex mechanisms are associated with costs that are not associated with simple mechanisms, then we would have an argument for a general cost-of-plasticity principle.

There are problems with this line of argument also. In Chapter 1 an example involving sex change in fish was used to suggest that functional complexity is not always associated with structural complexity. Sometimes a simpler device is more flexible than a complex one.

There is a more theoretical point to be made here also, concerning the concept of homeostasis. A homeostatic mechanism is one in which, when there is an environmental change, one set of organic properties also change, but their change has the functional consequence that another set of organic properties stays stable (section 3.4). One distinctive feature of mammals and other organisms which are usually thought to be structurally complex, when compared to simpler forms, is their greater use of homeostatic mechanisms. A central example is homeothermy, the maintenance of a constant internal temperature in the face of external changes.

Homeostatic mechanisms are complicated and expensive structures whose role is to bring it about so that the organism does something *simple* – their role is to bring it about so that some important set of variables changes as little as possible. Just as generating variety can be expensive and can require complicated underlying mechanisms, maintaining *stasis* can also be expensive, and can require mechanisms as paradigmatically complicated as those of mammalian physiology.

If left to its own devices, the environment will directly induce any amount of variation in an organism's states, simply as a consequence of causal links between the two. The environment will freeze, thaw, soak, desiccate, propel and pulverize the organism – it will send the organism on as complicated a tour of physical states as could be desired. A great deal of organic activity goes into preventing the environment from inducing such heterogeneity. Functional complexity is not the only thing which is costly in this context.

There is an obvious reply to this point. Organic heterogeneity in general might not always be more expensive than homogeneity, but we are concerned with heterogeneity that is adaptive. We are concerned with the production of organic variability that enables the organism to deal with its environment. Perhaps this type of complexity always has costs.

There are exceptions to that generalization as well. A case which is ideal for the present discussion is found in Richard Levins' *Evolution in Changing Environments* (1968a). Levins claims that the following pattern is common in invertebrates such as insects: high environmental temperatures speed up the development of the individual, compared to low temperatures, and the final result is a smaller adult body size. (This relation only holds within the range of temperatures in which more-or-less normal development is possible – temperatures too high will simply break the system.) This phenomenon, Levins says, seems to be a physiologically inevitable consequence of the animals' metabolic systems (1968a, pp. 13–14, 67–68).

This relation between size and temperature can have interesting consequences. Suppose that size in an animal such as a fruit fly is important with respect to retaining moisture and avoiding desiccation. When the environment is dry it is good to be somewhat larger than normal, as a larger individual has a proportionally smaller surface area and will not dry out as fast. But we have already noted that when it is hot, the individual will end up relatively small, as a consequence of basic metabolic factors. Then it is possible for these facts of metabolism either to work for the organism, or against it, depending on the structure of the environment. When they work for the organism, they make possible a free way of tracking the state of the world and producing a usefully variable response.

First, suppose that the hot areas in the flies' habitat also tend to be the humid ones, while the cooler areas are drier. Then the relation between temperature and body size works in the fly's favor. The fly will develop in such a way that it ends up larger in the dry environments, because they are cool, and it will be smaller in the moist environments, because they are warmer. Although it has not evolved a specific mechanism for doing this, it is able to track the state of the world with the aid of an environmental signal, temperature. The state of this signal modifies what the fly does, and this variable response adapts the fly to environmental variation with respect to humidity.

On the other hand, it may be that the warm areas are also the dry ones, and the cool areas are more humid. Then the metabolic relation between temperature and size works against the fly. In warm and dry areas, where the fly needs to be larger, it will develop into a smaller adult. There is still a correlation between the state of the fly and the state of the world, but if this is "tracking," it is tracking the fly could do without. A flexible pattern of response *maladapts* the fly to the pattern of variation in its environment.

Levins discusses two actual populations of fly, which may exemplify each of these alternatives. Flies from a population in the Middle East enjoy the free lunch, as in that environment humidity is correlated with heat. In contrast, flies around Puerto Rico, where dry areas are warmer, must contend with a natural antipathy between the basic facts of their metabolism and the structure of their environment.

The situation of these flies could be described with a mathematical apparatus like the one we have been using; the fly modifies its phenotype in response to a cue with certain reliability properties, and can produce the right or wrong phenotype for either wet or dry environments. The Middle Eastern flies have, without "paying" for a perceptual mechanism or other specialized devices, a reliable cue and an appropriate tuning of development to the state of the cue. These flies benefit from their own anti-homeostatic properties. The Puerto Rican flies also have a useful cue, but they are condemned to a worse than useless interpretation of it. It would be in their interests to evolve a more homeostatic, "canalized" developmental sequence, which does not attend to this environmental input, or to evolve a means for using the cue but responding in the other direction.

The point of this case is that the default assumption need not be organic constancy in the face of environmental variation, but rather tracking of that variation. When this is so, constancy or inflexibility, rather than flexibility, may incur a cost.[5]

We should now distinguish "arbitrary" from "non-arbitrary" relations between environmental stimulus and organic response. When an environmental signal is linked to the parts of the organism that respond to it by a long, complex causal path, any given state of the signal can be used to trigger a variety of possible responses, depending on which response is useful (Levins, 1968a). This is an "arbitrary" relation. Behavioral responses to perceptual states are mediated by attenuated paths of this kind, so what an animal does when it sees a red patch, for example, is not determined in advance and can be shaped evolutionarily and by individual learning.

The reaction of fruit flies to variation in heat is nonarbitrary. The pathway between the incoming signal and the internal structures which determine the individual's response is short. Temperature is not a signal transformed and filtered before it impinges upon metabolism; its effect is more direct. Then the range of possible metabolic responses to temperature is reduced. Note that temperature may have an arbitrary relation to other parts of the organism's response – it is not the signal itself which has this property.

A more short-term case of this type is the behavioral reaction to temperature by lizards and snakes. Whether or not a behavioral slow-down is a good thing when it is cold, this is what is produced, because the relation between ambient temperature and metabolic rate is nonarbitrary. In some circumstances, this nonarbitrary relation may give the organism a useful free mode of tracking the world. In other cases it might be the worst possible mode of response. In a large mammal, on the other hand, temperature changes are, within a certain range, arbitrary with respect to their effect on behavior. We can either speed up or slow down behavior when it is cold.[6]

Levins' example shows that there need be no intrinsic cost to having a usefully flexible strategy, a strategy which involves tracking the world and producing a response according to the state of an environmental signal. This case is not being given here as a model for mechanisms of plasticity in general. It might be very unusual. Further, it might now be suggested that it is not *useful* plasticity, but useful plasticity achieved through an *arbitrary* response, that is expensive. When a signal does not directly determine what the response must be, or when the nonarbitrary effects of the signal are bad, then there must be a mechanism for *effecting an interpretation* of the signal – such as an immune system, or a central nervous system. These are complicated and expensive, in terms of set-up and running costs.[7]

This last suggestion – that useful flexibility achieved through an arbitrary response is expensive – certainly has intuitive appeal. We are also reaching the point where a mass of empirical detail is probably needed to proceed further on this question. But I think there is reason to be cautious with this principle also. If the relation between signal and response is arbitrary, then *some* mechanism has to determine what interpretation is given. But it is a strong additional claim to say that these mechanisms are always associated with significant expense. Information and information-processing have become central commodities in modern life; they are bought and sold and occupy critical places in modern economies. We must be wary of extrapolating from our own case to that of sea moss and fruit flies. Sultan and Bazzaz (1993) found plastic responses to flood conditions latent in populations of a plant that had no apparent need for these responses. There was no indication that this plasticity was being lost in a situation where it was not needed – perhaps the cost is not significant. They also cite other cases in plants where no measurable cost is being paid for useful plastic responses.

So with this brief survey completed, no general principle about necessary costs to plasticity will be endorsed. Sometimes no mechanism is needed, sometimes the mechanism is cheap, and stasis, as well as change, can exert costs and demand complex underlying structures. However, this is not to say that maintaining a human brain or building a human visual system is cheap. In many particular cases, including the case of cognition, it appears that the mechanisms making flexibility possible require massive investments of materials and energy, bring about additional new risks of damage and breakdown, and greatly prolong development. The human brain, for example, consumes a large proportion of the total energy used by the body. Accordingly, it is worth investigating how consideration of some of these specific costs may influence a model of adaptive plasticity. In the next section we will look at the role played by one type of cost that is plausibly associated with some aspects of cognition – the cost of perceptual sampling of the world.

8.5   Paying for perception

This chapter has been concerned with ways in which an organism can shape the reliability properties of the cue it uses to track the state of the world. But this activity takes place within constraints imposed by the informational properties of the signal. The relation between the likelihood functions in Figure 8.1 exerts a constraint of this kind. The organism's setting of a threshold to determine its cue has no effect at all on the fact that an observation of a value of X around 5 is very ambiguous. When the two likelihood functions have a large area of overlap, the organism must contend with many ambiguous inputs. When the overlap is small the organism will more often receive high quality evidence of the state of the world. The overlap between the two functions has until now been treated as fixed in advance.

This last constraint will now be relaxed. The signal detection model can be augmented in such a way that questions about the costs and benefits of high and low degrees of perceptual acuity can be asked, and related to the cost-benefit considerations we have been using.

This will be understood as one way of building intrinsic costs of plasticity into the model. A cost can be imposed for each degree of perceptual acuity the organism acquires; the organism can pay a cost to reduce the overlap between the functions in Figure 8.1, thereby acquiring a more reliable signal. An inflexible organism pays no cost. We can describe a continuum between absolute inflexibility, flexibility which makes use of a poor environmental signal, and flexibility which has access to a highly reliable but costly signal.

One way to model this which is mathematically simple, although it admittedly sacrifices realism, is to understand perception as a process of sampling, where the cost imposed varies according to the size of the sample, as it would for a scientist. Every addition to the size of the sample improves the evidential value of this signal, but each addition to the size of the sample incurs a cost. The problem is to sample enough to get the reliability properties one needs, without paying so much for the sample that the improvement in reliability is a waste.

This picture will be grafted artificially onto the example with the bryozoans. We will think (hypothetically) of the bryozoan colony as perceiving the concentration of an informative chemical by some process of sampling. The cost paid by the bryozoan is proportional to the size (and hence the evidential value) of the sample. Perhaps the bryozoan has a fixed number of receptors whose individual observations are less reliable than an average value, because of micro-scale fluctuations in the concentration of the chemical or noise in the operation of receptors.

Figure 8.1 is then interpreted as representing the distribution of "average" values when the sample size is one. This is the cheapest and least reliable way to sample the water.

Suppose now that there are two receptors, whose output is averaged. At this point we can make use of a convenient statistical fact. The distribution of values of the mean of a sample from a normal distribution, is normally distributed, and the standard deviation (or "spread") of the distribution of sample means is related in a simple way to the standard deviation of the original distribution from which the sample is taken. The larger the sample, the smaller the standard deviation of the distribution of observed sample means.[8] The upshot is that if we are observing an average of a sample of size two, the likelihood functions associated with our observations are not as they were in Figure 8.1. They are the same shape but they are more sharply peaked and have less overlap between them.[9]

Suppose these new functions are used and the rest of the model is retained intact. Then the optimal threshold value of X is at $X = 5.27$. Parameters a_1 and a_2 are 0.96, and 0.85 respectively. This improvement in reliability is reflected in the overall expected payoffs associated with the strategies. When the organism has one receptor only, its expected payoff, using the optimal threshold, is about 9.30. When it has two receptors (and switches to the new optimal threshold) its expected payoff is about 9.44. As the organism gains receptors it gains reliability, and this is reflected in a higher expected payoff.

As we might expect, there is also a "diminishing return" effect which rapidly reduces the benefit associated with adding receptors and improving accuracy. And as statisticians accept that after a certain point taking additional observations is not worth the effort, achieving an extreme level of reliability in one's tracking of the world may be self-defeating. Indeed, if the cost of having a receptor is high when compared with the importances of the states of the world, it may be best even in the presence of good signals to have no receptors at all, and fall back to an inflexible strategy.

A range of familiar economic considerations is relevant here. In this model, payoff increases with the number of receptors used to track the world but it does so in a decelerating way. The cost of adding receptors might increase in the same way or it may increase differently. Considerations such as these will determine how the optimal trade-off between the cost of sampling and its advantages is achieved.

To take a very simple example, suppose the cost of adding and maintaining receptors increases linearly from zero by 0.1 of a unit of fitness per receptor. Having one receptor incurs a cost of 0.1, having two costs 0.2, and so on. If

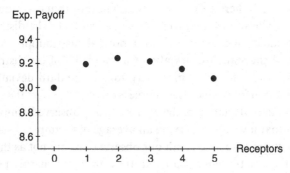

Figure 8.2. Overall payoff as a function of receptor number

246

this is a "per decision" cost, these costs can be subtracted from the expected payoffs associated with each level of reliability, resulting in an overall or net expected payoff. These net payoffs are represented in Figure 8.2. The best policy for the hypothetical bryozoan colony in this situation would be to have two receptors and no more. The increased payoff from higher accuracy is not worth the cost.

The details of particular cases are not the issue here. The point is to illustrate how the benefits of reliability as a property can be traded off with costs that may be associated with its underlying mechanisms.

8.6 On reliability

In this chapter a decision-theoretic framework has been used to investigate properties of reliability. In this section I will try to establish some links between this decision-theoretic treatment and some other discussions of reliability in recent epistemology.

In the models of this chapter and the previous one, reliability properties were represented with parameters a_1 and a_2. Before trying to draw epistemological morals from the models discussed, it is important to note several differences between the way reliability has been handled in these chapters and the way it is handled in other discussions.

First, parameters a_1 and a_2 were associated with phenotypes or behaviors, not with beliefs or perceptual states. Reliability was treated decision-theoretically by taking it to be a relation between behaviors and conditions in the world which determine the success of behaviors. Most treatments of reliability in recent epistemology have not had this orientation. They have investigated the reliability properties of beliefs or other inner states which are considered independently of their links to particular behaviors. The criteria for success have not been behavior-linked, but derived from common-sense standards and intuitions.

Second, the concept of reliability I have been using is *two-dimensional*. An agent is evaluated with respect to a_1 and a_2, and one cannot be reduced to or expressed in terms of the other.

Third, the measures a_1 and a_2 in the models are conditional probabilities of the form $Pr(C_i|S_i)$. The state of the world is the condition; the probability expresses how likely it is, *given* a state of the world, that the agent will produce an appropriate behavior. This does not tell us how likely it is, *given* that an agent has made a certain choice, that the choice will be right. That would be $Pr(S_i|C_i)$.

247

In contrast, the concepts of reliability used most often in reliabilist epistemology are one-dimensional, and they have to do with the probability of a state of the world given a state of the agent.

This is not always obvious, as not all reliabilists make overt use of probability, and there are also some views that are exceptions.[10] But the most familiar way to make use of reliability in recent epistemology is to say: given that this belief or other inner state has been tokened in an agent, how likely is it to be true? We start from the existence of an inner state of a certain type – perhaps an inner state that is the product of a specific perceptual or psychological process – and we ask what the existence of the inner state guarantees or makes likely about the state of the world.

An example is Armstrong's theory (1973, pp. 166–68). Armstrong says that a person has non-inferential knowledge that p if there is a specification of the condition and circumstances of that person such that: if he believes that p, then it must (nomically) be the case that p. This is not expressed in terms of a probability, but it does imply one, if probabilities can be applied to the situation. It implies that the probability that p, given that p is believed, is one.

The theory of Dretske (1981) does use probability explicitly. Dretske identifies knowledge with information-caused belief. You know that p if your belief that p was caused or is causally sustained by the information that p. A signal carries the information that p if, given the state of the signal, the probability that p is one, though it would not be one independent of the state of the signal. In Dretske's account it is the signal which causes the belief, not the belief itself, that has the reliability property. But the reliability property is of the same type: it describes how likely a distal state of the world is, given the state of the signal.

Goldman (1986) uses reliability in analyzing both knowledge and justification. He does not appeal to probabilities or to subjunctives. A reliable belief-forming process is simply one which produces a high proportion of true beliefs among the beliefs it produces (1986, p. 26). To qualify as knowledge, a belief must be (among other things) the causal product of a reliable belief-forming process. So again, if probabilities can be applied at all to the situation we can say: given that a belief has been produced by this belief-forming process, it has a high probability of being true.

Many reliabilist theories are concerned primarily with the analysis of common-sense concepts of knowledge and justification. My aim, on the other hand, is to look at ways to use reliability concepts in giving a naturalistic account of epistemic value. But there is one feature of the way reliability was used in the biological models that can be applied more generally in epistemology: reliability-based epistemic assessment should be *two-dimensional* .

This idea can be motivated with the aid of an informal argument which has been used by a variety of writers. Consider an agent so pathologically cautious in belief-fixation that he only accepts a proposition if he has the best possible evidence for it. The beliefs he does end up with will have excellent epistemic properties – almost all true, perhaps. But this stock of beliefs will be extremely small. This agent has achieved reliability, but only through missing out on most of what is going on (James, 1896; Levi, 1967; Rescher, 1977; Goldman, 1986).

As most reliabilists have been concerned with the analysis of specific common sense concepts, it is not a criticism of them to say that they have not given a complete account of epistemic value. But by making changes to the existing framework of reliability-based epistemic assessment, we can develop a richer apparatus and come closer to a complete account. A complete account of epistemic virtue needs to take into account both how likely an agent's beliefs are to be true, and how likely he or she is to come to believe a given truth.

Of the reliabilists mentioned above, Goldman is the most concerned to take account of this issue. He regards reliability as one virtue, but also recognizes another distinct virtue, "power," which is the ability of a belief-forming process to deliver true beliefs in response to a large proportion of the problems the agent wants to solve (1986, p. 27). An alternative way to approach this issue is to modify the concept of reliability, and view it in a two-dimensional way.

On the view I am proposing, there are two different questions we can ask about an agent's "reliability" with respect to some belief or inner state. We can ask how likely that belief is to be true, given that it is tokened, and we can ask how likely it is to be tokened, given that it is true. The second type of question concerns an analog of Goldman's "power," though it is not the same concept. This second property is also sometimes called "sensitivity." I will refer to both properties as "reliability" properties in order to stress some properties of the relationship between them, and to stress the fact that they can be discussed simultaneously in the same probabilistic framework.

We will take B_i to be an inner belief-like state of the type discussed in reliabilist epistemologies. S_i is a state of the world which this inner state represents or has as a truth-condition. The reliability measures standard in reliabilist epistemology describe values such as $Pr(S_i|B_i)$; they describe the probability of the world being the way the agent thinks it is. The other measurement starts from the state of the world, and asks how likely the agent is to token some inner state in response: $Pr(B_i|S_i)$. This is the type of measure we have been using in the models. Both measures are to be understood in terms of an objective or physical conception of probability.

There is an important mathematical relation between the two probabilities. They are related via Bayes' theorem, and it is a consequence of Bayes' theorem that correlation (in the sense of positive relevance) is a symmetrical relation. If $Pr(S_i|B_i) > Pr(S_i)$, then $Pr(B_i|S_i) > Pr(B_i)$. This fact can lead one to think that "having thoughts correlated with the state of the world" is a single property, or a single epistemic goal. If your belief that p makes it more likely that p is true, then p being true makes it more likely that you will believe p. That symmetry does exist, but it is *not* the case that "having thoughts correlated with the state of the world" is a single epistemic goal. Having a high value of one type of reliability, say $Pr(S_i|B_i)$, is not just a different property from having a high value of $Pr(B_i|S_i)$. In many cases it is impossible to pursue them both at the same time. One must be traded off against the other.

This can be put informally, as I did above and as William James did very forcefully (1896): studiously avoiding false belief is not always a good way to make oneself likely to acquire a large stock of true beliefs. Maintaining a large stock of commitments, having a strong and detailed view of how the world is, requires gambling with the possibility of error.

If the choice is between believing that p and believing that not-p, with no possible middle ground, then the trade-off does not arise. Though p and not-p may differ in their level of content, if one must make a commitment either way then avoiding falsehood is the same as seeking true belief. The trade-off is also avoided if the agent does not make qualitative commitments to propositions, but has a stock of propositions and maintains only a *degree* of belief for each of them, which is updated in the light of new evidence (Jeffrey, 1983). The situation where the Jamesian trade-off applies is where the choice is between accepting a belief and "doing without it" (James) or "suspending judgement" (Levi).[11] In that case, addition of a new belief to one's stock may well be acquiring a truth, but will generally also be increasing the risk of error.[12] So an agent can be "reliable" in one sense when whatever is in her head is a good guide to the state of the world; her beliefs tend to be true. But an agent can also be "reliable" in another sense when there is a large range of conditions in the world which she is a source of information about. An epistemically cautious hermit is reliable in the first sense, but not in the second sense.

As I said before, some prefer not to use the term "reliability" for this second sense, but would use "sensitivity" or some other term. My point here is not conceptual analysis, and it is less important to have the right name for this second property than it is to have a place for it in the theory. But I will call them both reliability properties. Each property will be named after a philosopher who has cared about it. The measure $Pr(S_i|B_i)$, the more familiar one, will be called *Cartesian Reliability*; the other, $Pr(B_i|S_i)$, will be called *Jamesian Reliability*.[13]

Using these names for the two properties should not be taken to suggest that Descartes did not care about what I am calling Jamesian reliability and that James did not care about the other property. Neither of those claims is true. But each philosopher is famous for epistemological theorizing which focused on a particular one of these properties. Descartes' *Meditations* sought to establish and meet high standards with respect to Cartesian reliability, and James wrote "The Will to Believe" with the aim of defending the importance of Jamesian reliability. Indeed, a central point in James' discussion is the need to recognize and deal with a necessary trade-off between the two goals of thought.

The trade-off James described can also be discussed more formally. One way to do this is to make use of the basic framework of the signal detection model, while modifying the interpretation of key concepts within it. Suppose there are two states of the world S_1 and S_2, as before. The agent can either token B_2, which has S_2 as truth condition, or she can decline to token it ("do without it"). Doing without it may include disbelieving it, suspension of judgement, or perhaps also behaving in such a way that the issue does not arise. We are concerned with when the agent will make the positive decision to "take on board" B_2. A more complete account would treat all these options separately (Levi 1967, 1981), but one aspect of the problem can be illustrated in this simplified setting.

The agent has access to an environmental signal represented in Figure 8.1. The decision, as before, is where to put the threshold, the point at which evidence is regarded as good enough for the agent to token B_2. It is assumed that the agent makes a qualitative decision, rather than adjusting a continuous degree of belief. If we had a way to determine the "epistemic utility" of the various kinds of right and wrong decisions, and also a way to interpret all the relevant probabilities, then we could set an optimal threshold. The "importance" of a state of the world would be the difference in epistemic utility between believing it true and not believing it to be true, when it is true. The formulas of the signal detection theory apparatus can be applied as before. I will not discuss how the utilities might be assigned here. The present point has to do with the basic nature of the trade-off between Cartesian and Jamesian reliability.

The location of the threshold determines how the trade-off between the two types of reliability is achieved. The parameter a_2 from the earlier models is the same as Jamesian reliability. Jamesian reliability is pursued by setting the threshold towards the left; even relatively low values of X are regarded as good enough evidence to prompt tokening of B_2. The agent is unlikely to fail to token B_2 when S_2 is the case. But if the threshold is over to the left the Cartesian reliability is reduced. There is a good chance that B_2 will be tokened even though S_1 is the actual state of the world. High Cartesian reliability is sought by moving the threshold to the right, but this entails a loss in Jamesian reliability.

This trade-off will not apply in every situation. It applies here because every time the agent adds a new type of evidence to the set which is regarded as good enough to induce tokening B_2, a *worse* grade of evidence is being added. If an agent adds a piece of evidence to the set which is better than some of what is in there already, both the Cartesian and Jamesian reliabilities can be improved. The problem modeled here is that of deciding what *quality* of evidence to demand, when making an epistemic decision. This decision will be made by a rational agent in accordance with his or her goals and values; there is no grade of evidence which is intrinsically good enough or not good enough to demand acceptance, for any agent at all. The trade-off between the two kinds of reliability is a decision-theoretic problem.

As mentioned above, the trade-off is avoided if the agent does not make qualitative commitments to propositions, but rather maintains and updates a degree of belief. This is the conception of belief usually used in decision-theoretic approaches to evidence and learning. From this point of view it can seem artificial to impose qualitative choices on agents with respect to epistemic matters; *behavior* may often involve a qualitative choice, but underlying behavior there can be a purely quantitative property of degree of belief. This is an important problem with the attempt to use the signal detection framework, and its ilk, to model purely epistemic decisions, and it is not taken lightly here. The underlying aim of this discussion is to introduce concrete relations between inner states and the world into the decision-theoretic approach; the aim is to motivate developing a decision-theoretic framework in a less internalist way. An attractive way to do this is with reliability properties, and the simplest way to make use of reliability properties here is to understand them as links between beliefs and truth conditions. But there is a need to undertake a more systematic fusion of the reliabilist and decision-theoretic frameworks.[14]

In sum, there is good reason to think that a two-dimensional concept of reliability, such as the concept used here, will enable reliabilism to give a more complete and integrated picture of epistemic virtue. Reliabilism in general is on the right track, in its attempt to investigate epistemic value by attending to physical relations between internal states of agents and conditions in the agents' environments. What is needed is an integration of these traditional reliabilist concerns with factors described in pragmatist and decision-theoretic epistemologies.

Notes

1 A standard reference on signal detection theory is Green and Swets (1966). Another good introduction is in Coombs, Dawes and Tversky (1970). In Godfrey-Smith (1991) I give a simplified presentation of the basic apparatus. That version uses only discrete variables and hence no calculus. Some readers might find the 1991 version a useful introduction to this one.

   See also Cohen (1967) for another discussion of information and environmental variability which stresses many of the themes of this chapter and the previous one.

2 There is no reason why the likelihood ratio should have one function rather than another on the top. I am trying to keep the language simple.

3 This trade-off between error probabilities is familiar in the part of statistics that deals with hypothesis testing. If we regard S_1 as a "null hypothesis," then $(1 - a_1)$ is the probability of a Type I error, and $(1 - a_2)$ is the probability of a Type II error (Neyman, 1950). Statisticians stress that the two types of error cannot be minimized simultaneously, within the constraint of a fixed sample size (Kendall and Stuart, 1961, p. 183). This terminology can be useful, but the term "null hypothesis" is a loaded one which must be treated with caution.

4 MacArthur (1972) and Levins (1968a) are often cited as the works behind much modern application of Jack of all Trades principles in ecology and evolutionary biology. The idea is much older though. For a historical discussion see Sultan (1992).

5 See Tierney (1986) for an interesting discussion of some of these issues, stressing neuroscientific considerations. Tierney argues that plasticity and flexibility, rather than inflexibility, should be viewed as the primitive state or starting point in the evolution of nervous systems. Inflexibility or canalization of behavior evolves from flexibility.

6 I have been using this case from Levins as a counter-example to some intuitive ideas about necessary costs of plasticity. It is interesting that Levins himself, in his 1968 book, does nonetheless accept a strong principle about costs: "Since survival in a variable environment involves a homeostatic system which imposes a cost, this will reduce fitness in the constant environment" (p. 102). Levins may have in mind homeostasis involving "arbitrary" responses.

7 I interpret G. C. Williams' distinction between "response" and "susceptibility" as being at least very close to this distinction between "arbitrary" and "non-arbitrary" responses (Williams, 1966, Chapter 3).

8 If $\sigma(\bar{X}_N)$ is the standard deviation of samples of size N, and σ_X is the standard deviation of the original population, then $\sigma(\bar{X}_N) = \sigma_X/\sqrt{N}$. The standard deviation of the original distributions (Figure 8.1) was 1, so if the sample is of size 2 then the standard deviation of each new distribution is 0.71.

9 The evidential value of an environmental signal can be measured with a quantity called d'. The quality of the pair of functions in Figure 8.1 is measured by finding the difference between the two means of the distributions (μ_1 and μ_2), and dividing this difference by the common standard deviation of the distributions.

(21) $d' = (\mu_1 - \mu_2)/\sigma_X$

In Figure 8.1 the distributions both have standard deviations of 1. So $d' = 2$. If d' is high it is easy to find a cue with good values of a_1 and a_2. The d' value with two receptors is about 2.83.

10 An exception to some of what I say about reliabilism is the view of Nozick (1981). Nozick's concept of "tracking" the world is a two-way concept. He does not count a belief as knowledge simply if it has to be true, given that it is tokened. He requires also that it has to be tokened, given that it is true. All the reliabilist theories mentioned here, including Nozick's, are discussed in more detail, and related to a decision-theoretic framework, in Godfrey-Smith (unpublished).

11 It is central to James' "will to believe" argument that doing without a belief can, in some cases, have the same *practical* upshot as disbelieving it. James need not say that there is no *cognitive* difference between doing without and disbelieving.

12 The most detailed models of this type of epistemological problem have been developed by Levi (1967, 1981).

13 In earlier presentations (1991, 1992) I have used terms borrowed from Hartry Field (1990): $Pr(S_i|C_i)$ was called "head-world reliability" and $Pr(C_i|S_i)$ was "world-head reliability." But the associations with Descartes and James should have better mnemonic properties.

14 There may be ways to recast some of the issues discussed in this chapter in a Bayesian, degree-of-belief framework. The agent's degree of belief might be treated as something like an estimate of the truth-value of a proposition. True propositions have value 1 and false propositions have value 0. An agent with a subjective probability of 0.9 in a true proposition has a better estimate of that proposition's truth value than an agent with a subjective probability of 0.1. Then an agent could be assessed with respect to the average accuracy of their degrees of belief; we could measure the average deviation of their degrees of belief from the beliefs' actual truth-values.

9

Complex Individuals, Complex Populations

9.1 Another kind of complexity

In Chapters 7 and 8 we looked at the situation of an individual faced with a complex environment, and described some conditions under which it is best for the individual to meet this environmental complexity with a complex organic response. This is not the only type of organic system which can be complex or simple; another is the *population*. This chapter is about population-level responses to environmental complexity, and the relation between individual-level and population-level responses.

Complexity in the case of individuals was understood as heterogeneity. Complexity in populations will be understood the same way. A complex population is one which contains a diversity of types or forms. A simple population is homogeneous. The models of Chapters 7 and 8 examined a particular kind of individual-level complexity: the ability of an individual to do a variety of different things in different circumstances. In this chapter we will look at a particular case of population-level complexity: heterogeneity or "polymorphism" in the genetic make-up of the population. We will also investigate the relations between two different realizations of biological complexity, the relations between (i) simple populations of complex individuals, and (ii) complex populations of simple individuals.

The first part of this chapter is only indirectly relevant to the environmental complexity thesis. The aim is an understanding of a pattern of externalist explanation of complexity in biology, and certain work on genetic polymorphism will be discussed as a case study.[1] In Chapter 2 some ideas in population genetics were discussed as expressions of an internalist explanatory strategy. When explaining the course of evolution, or the observed characteristics of plants and animals, genetic factors are internal. In this chapter, on the other hand, a genetic property is the *explanandum*, that which is being explained. We will look at explanations of genetic heterogeneity in terms of environmental heterogeneity.

The aim of this discussion is not to *assert* a c-externalist principle about genetic heterogeneity, but to look at how c-externalist programs in this area have been developed, and to attempt to clarify some conceptual issues surrounding these explanations.

Later we will also reach a point of direct contact with the discussions of mind and environmental complexity in earlier chapters. We will look at the possibility of formulating principles which describe when we should expect to find individual level complexity rather than population level complexity. Are there particular types of environmental variation that tend to generate one or the other? In this context it is often suggested that a rapid pace of environmental change makes individual adaptation necessary, while slower changes can be tracked genetically (Bradshaw, 1965). This is an intuitive idea and may be correct. In this chapter however we will look cautiously at some more abstract relationships between complexity in individuals and populations. Arguments which modify the work of Richard Levins (1968a) suggest that some properties of geometric mean fitness measures, introduced in Chapter 7, provide a general reason to expect individual-level rather than population-level complexity in certain restricted circumstances.

9.2 Polymorphism

The property of population-level complexity we will focus on is *genetic polymorphism*. A population is polymorphic with respect to some locus (position on a chromosome) if there is more than one allele (alternative form of a gene) which individuals can have at that locus. A polymorphic population is more complex or heterogeneous, in this respect, than a population which contains only one allele at that locus.

We will only be concerned with evolutionary processes occurring at one locus, and usually involving two alleles. Most of the models we will discuss apply more generally, but we will look only at the simplest cases. The alleles are labeled A and a.

We must distinguish between two types of population. The first consists of individuals which reproduce asexually and are *haploid*. These individuals have only one chromosome of each type, and hence only one allele at the locus in question. Offspring of these individuals are clones of the parent, except for occasional mutations. (Though we will consider only asexual haploids, there can also be sexual reproduction in haploids.)

The other type of population reproduces sexually and is diploid. A diploid individual has two sets of chromosomes, one from each parent, and hence has two copies of each gene. A diploid, sexual individual passes only one of these two copies to its offspring when it reproduces. The two copies possessed by an individual may or may not be of the same allele. An individual with two copies of the same allele is a homozygote, symbolized *AA* or *aa* as the case may be. An individual with one copy of each allele is a heterozygote, symbolized *Aa*. So in a population of diploids, when there are two alleles at a locus there are three genotypes: *AA, Aa, aa*. In a haploid population there are only the genotypes *A* and *a*. (There can be asexual reproduction in diploids but we will not consider this case either.)

There are various possible relationships between genotypes and phenotypes, the structural and behavioral forms that individuals display. Each genotype may have its own associated phenotype. Alternately, the phenotypes of *AA* and *Aa* can be identical. Then allele *A* is *dominant* to *a*, and *a* is *recessive*.

An individual's phenotype determines its fitness. Fitness is understood as expected relative reproductive output. The fitness of *AA* is represented as W_{AA}, that of *Aa* as W_{Aa}, and so on. In simple systems, these genotype fitnesses, along with the frequencies of genotypes, suffice to predict the course of evolutionary change from generation to generation.[2]

Even in systems as simple as this, it cannot be assumed that the "fittest" will always "survive" and spread through the population. For example, suppose the fittest genotype is *aa*, but allele *a* has only just entered the population and is rare. Then the fate of allele *a* will depend on the fitness of *Aa*, as *a* alleles are almost always found in heterozygotes when *a* is rare. If *Aa* has a lower fitness than *AA*, then allele *a* will be lost, even though it would do very well if it became common enough to be seen in *aa* homozygotes.

A population is polymorphic at a locus if there are at least two alleles at appreciable frequency at that locus. We will look at several selective mechanisms which will maintain polymorphism, starting with the simplest and most famous.[3]

Suppose genotype fitnesses are constant. Then a stable, "balanced" polymorphism can exist if the heterozygote *Aa* has the highest fitness of the three. This is called *overdominance*, or *heterosis*. The latter term has some ambiguity associated with it, being used sometimes to refer to the general phenomenon of the superiority of cross-bred individuals (Gowen, 1964), and sometimes to an underlying mechanism or cause – the superiority of the heterozygous state at particular loci. I will use it here to refer just to this pattern of fitness relationships – what Lewontin (1974a) calls "single-locus heterosis," although some writers I will discuss, such as Dobzhansky, used it in the more general way.

This mechanism provides the only way there can be stable polymorphism maintained by selection on one locus with constant fitnesses. As a result there is the maintenance of variability; at equilibrium, the population contains a diversity of genotypes and phenotypes.[4] Heterosis can only exist in a diploid population. In a haploid population there are no heterozygotes.

Questions about the importance of heterosis as an evolutionary factor have occupied a pivotal place in the history of genetics. In particular, a central concern of population genetics between the 1940's and 1960's was a dispute about the amount of genetic diversity in natural populations, and the mechanisms which maintain this diversity. The contending positions have been labelled the "classical" and "balance" views (Dobzhansky, 1955; Lewontin, 1974a). The classical view associated with H. J. Müller, held that the fittest type is generally a homozygote, and variation introduced by mutation is largely a deleterious "load" carried by a population, which is constantly being weeded out by selection. On the balance view, associated with T. Dobzhansky, genetic variation is ubiquitous, at least in many populations, and it is maintained in these populations by the operation of selection.

The issue of the prevalence of heterosis in nature was important in this debate. Heterosis is not the only way in which selection can maintain variation, but it is the theoretically simplest way, and a large number of experiments were conducted to investigate the phenomenon (Dobzhansky et al., 1981). Accompanying this experimental work was theoretical investigation directed at (i) saying *why* it might be that heterozygotes were fitter, and (ii) describing new ways in which selection could maintain genetic variation in a population. Some of this theoretical work fits squarely into an explanatory schema we are concerned with in this book: it explains organic complexity in terms of environmental complexity.

9.3 Individual homeostasis

The maintenance of genetic variation is a consequence of heterosis; what might be a cause of this pattern of fitnesses?

First, let us note that the idea that heterozygotes might be fitter than homozygotes has a long experimental prehistory (see Paul, 1992b). It was a familiar observation before genetics that hybrids or cross-breeds often have higher values of various phenotypic properties – they are often larger, more robust and so on. The phenomenon was called "hybrid vigor," and once a genetic framework was developed it was a straightforward observation that hybrids will tend to be heterozygous for many genes.

We should also note that there are various explanations which can be given for heterosis that do not involve environmental complexity. Heterozygotes, even when they have intermediate values of some phenotypic properties, might thereby have higher values of other phenotypic properties, and/or higher fitness. They may tend to strike a "golden mean" between too little and too much of an enzyme, for example. There is also the famous case of sickle cell anemia. Here the heterozygote produces a partially "sickled" hemoglobin molecule which does not lead to serious anemia and which also provides resistance to malaria that the normal homozygote does not. The heterozygote is fittest overall, but not because of its property of complexity.[5]

However, environmental complexity played an important role in debates about heterosis. I will look at two explanations proposed which were c-externalist in form.

Dobzhansky believed that heterozygote individuals are more *homeostatic* than homozygotes. Heterozygotes are better able to adjust to environmental changes in such a way as to "maintain their internal milieu in functional order" (Dobzhansky and Wallace, 1953, p. 168. See also Dobzhansky and Levene, 1955). In specifying what homeostasis is, Dobzhansky (1955) appealed to Cannon's (1932) account in terms of physiological "steady states." He claimed that functional homeostasis of the type described by Cannon would produce "developmental homeostasis" (1955, p. 7). But Dobzhansky's genetic work inevitably went beyond this conception, and in practice he applied the concept of homeostasis in a more informal way, as involving adaptive stasis in the face of environmental variation. We should note also that Dobzhansky primarily studied individuals heterozygous for whole chromosomes, or whole stretches of chromosome, not just a single locus.

Suppose heterozygotes are more homeostatic. So far this is not a c-externalist explanation. But it is now possible to ask: why are they more homeostatic? Here we need to look more closely at what heterozygotes are. If an individual has two different alleles at a locus, both these stretches of DNA are usually transcribed and give rise to a protein. Consequently, a heterozygote can produce two proteins and a homozygote can only produce one. This may give a heterozygote two available forms of an enzyme, a chemical which controls reactions between other chemicals. Thus there is a respect in which a heterozygote is individually more complex than a homozygote; it has genetic variability of its own. This last step in the explanation was taken by F. W. Robertson and E. C. R. Reeve (1952).

> The more heterozygous individuals will carry a greater diversity of alleles, and these are likely to endow them with a greater biochemical diversity in development. This will lead to heterosis, because of the more efficient use of the materials available in the environment, and also to a reduced sensitivity to environmental variation, since there will be more ways of overcoming the obstacles which such variations put in the way of normal development. (Robertson and Reeve, 1952, p. 286)[6]

This explanation, directed ultimately at explaining population-level diversity, is not only based upon the properties of individuals, but is based specifically upon individual adaptability. The problem of complexity in a population is addressed by (i) focusing on the case in which this complexity appears in an individual, and then (ii) giving a c-externalist explanation for the advantage held by this type of individual.

If this view is adhered to strictly, the existence of genotypes *AA* and *aa* in a sexual population is seen as an incidental consequence of the fact that heterozygotes do not breed true. Polymorphism is not a response to environmental complexity distinct from individual-level response, but is a byproduct of it. The existence of population-level complexity is a consequence of the advantages of individual-level complexity, along with the facts of sexual reproduction.

This basic view of the maintenance of population-level complexity was expressed also by I. M. Lerner in his book *Genetic Homeostasis* (1954). Lerner accepted that heterozygotes are generally more homeostatic than homozygotes, and this gives them a selective advantage. He thought the "buffering" displayed by heterozygotes should be explained in cybernetic terms. He did not commit himself to the idea that the sheer "biochemical diversity" in heterozygote individuals is the key to this, as claimed in the quote above from Robertson and Reeve. That is one possibility, but another is that heterozygotes tend to produce an appropriate intermediate level of certain chemicals (pp. 65–66); this is a "golden mean" principle rather than a c-externalist principle. Lerner raises other possibilities as well. The upshot is that polymorphism, which is caused by individual homeostasis, in turn generates population-level plasticity, which Lerner called "genetic homeostasis." This population-level plasticity is a "by-product" or "after-effect" of the fitness properties of heterozygote individuals (1954, pp. 118, 120).

The explanation of polymorphism in terms of the homeostasis of hetero-zygotes has received less attention in recent years. One interesting recent discussion, however, is that of Orzack (1985). Orzack agrees that there is experimental support for the idea that heterozygotes are more homeostatic, as individuals, than homozygotes. He claims, however, that there is no evidence that homeostasis is a globally beneficial biological property. Neither organic stability in general, nor stability with respect to fitness properties, can be generally assumed to be beneficial. He presents a mathematical model in-tended to show that more homeostatic genotypes can be either favored or not, depending on the specificities of the situation. His conclusions are strong: "The notion of homeostasis as a natural and intrinsically beneficial attribute of organisms has insinuated itself into many areas of evolutionary biology. It is my belief that this concept rarely serves to enlighten and should be discarded for lack of a precise definition" (1985, p. 570).[7] I agree with Orzack that a tendency to use an imprecise concept of homeostasis, associated in a loose way with stasis or reduced variance, has led to problems in this area. We saw in Chapter 3 that Lewontin's reaction to these problems was to define homeo-stasis simply in terms of survival in a large range of environments. But there is also an alternative response to Orzack's and Lewontin's. This is to (i) develop a precise and narrower concept of homeostasis, and (ii) resist the temptation to assume that homeostasis is always progressive, or the best option.

In Chapter 3 a specific and narrow definition of homeostasis was given. If O_1 and O_2 are different organic variables, a homeostatic pattern of response is one for which the following conditions are true: survival is improved by stasis in O_1, no matter what O_2 does. It is not the case that survival is improved by variation in O_2, no matter what O_1 does. But O_2 does affect survival, as variation in O_2 is positively correlated with stasis in O_1. This concept of homeostasis does have a conceptual connection with an improvement in survival. But it is not assumed that homeostasis in this sense is always worth the associated costs, and always better than alternative ways of dealing with the environment.

In this sense of homeostasis, Robertson and Reeve's explanation of hetero-sis in terms of "biochemical diversity" involves genuine homeostasis. But an explanation of the superiority of heterozygotes in terms of their intermediate phenotypic properties, their striking a "golden mean," is not homeostatic. Neither is sheer developmental or physiological stasis in the face of environ-mental change. Some of Lerner's and Dobzhansky's uses of "homeostasis" do not fall within this narrow conception; not all stasis is homeostatic, and not all adaptive stasis is homeostatic.

261

9.4 Homeostasis and the population

We will now look at a second explanation proposed for the phenomenon of heterosis. There are suggestions of this idea in Dobzhansky's work (see Dobzhansky and Spassky, 1944; Dobzhansky, 1951, p. 110), but it was elaborated in more detail by others. This second explanation is also c-externalist, but it relates genetic variation in the population directly to environmental variation, without going via the individual-level properties of heterozygotes. It is, in fact, a mirror image of the causal structure of the first explanation.

Dobzhansky and others saw a fundamental symmetry between individual-level and population-level responses to environmental variation.

> Adaptation to a variety of environments is accomplished in two ways. First, most species and populations are polymorphic and consist of a variety of phenotypes optimally adapted to different aspects and sequences of environments. Secondly, individuals respond to environmental changes by physiological and structural modifications. (Dobzhansky and Wallace, 1953, p. 162)

The second explanation for heterosis claims that the population as a whole benefits from being genetically diverse. Genetic diversity is brought about by heterotic fitness schemes, and preserving this diversity is the function of heterosis.

On this view heterosis is a population-level adaptation. A strong version of this position was defended in the 1950's by Lewontin (1956, 1957, 1958). The idea is that for any environmental condition encountered, a polymorphic population is more likely to contain some individuals which can survive in it, than in a homogeneous population. If some individuals can survive, the population can survive. Lewontin claimed that the explanation for heterosis is its generation of polymorphism which benefits the population as a whole. "Polymorphism is a homeostatic device by which populations can persist despite fluctuation in environment" (1958, p. 502). It is not that heterotic fitnesses are more likely to exist, in the abstract, but these are the schemes of fitness we are likely to observe, as these are the schemes of fitness that characterize populations which are likely to survive.

As Lewontin said explicitly, this requires a process of group selection.

> In a variable environment a polymorphic population is more highly adapted than a monomorphic one and the evolution of polymorphism from monomorphism is a product of *interpopulational* rather than *intrapopulational* selection. (1957, p. 398)

The difference between this explanation and the one discussed in the previous section is sharp. In the first view, polymorphism in a *population* is a side-consequence of the homeostatic properties of heterozygote *individuals*, and the explanation for the superiority of these individuals is c-externalist. On this second view, the superiority of the heterozygote *individual* is a side-consequence of interpopulation selection for polymorphic *populations*, where the explanation for the superiority of polymorphic populations is c-externalist. The two explanations are precise causal inversions of each other, though both are c-externalist.

Lewontin in fact presented the individualist and group selectionist alternatives in such stark contrast that he established a false dichotomy.

> One must either argue that heterozygosity *per se* confers greater fitness on an organism because of some biochemical mechanism basic to all living matter, or else that the higher adaptive value of heterozygotes is itself a product of evolution. (1958, p. 494)

This is a false dichotomy because one individualist alternative, which we have discussed, holds not that heterozygotes are always better as a consequence of basic properties of all life, but rather that heterozygotes are always better *in a variable environment*. An alternative to the idea that heterosis evolved from other fitness schemes is the c-externalist idea that complex individuals always do better in a complex world. (So the experiment Lewontin used in this 1958 paper – which showed loss of heterosis in a *stable* environment – does not tell against the c-externalist version of the individualist view.)

In 1966 G. C. Williams published his *Adaptation and Natural Selection*, a book which, along with other work around this time, produced a considerable individualist turn in evolutionary thinking (see also Hamilton, 1964; Lewontin, 1970). Williams' central target was biological explanation in terms of "the good of the species." However, many found all informal attributions of properties of adaptive organization to populations and other higher-level entities suspect, from this point. Subsequent investigation of evolutionary mechanisms also had to take heed of a series of models which purported to show that group selection was bound to be a weak force, in comparison with individual selection.[8]

The development of a strict distinction between individual and group selection, and the disrepute into which group selection fell, caused problems for the idea that properties of genetic systems can be explained in terms of selection between populations. However, another line of theoretical investigation into polymorphism had arisen in the 1950's as well. This approach was not troubled by the individualist turn in evolutionary theory; although it is another example of c-externalist explanation at the population level, the mechanism used is individual selection. This approach has become a central part of the project of giving selectionist explanations of polymorphism.

9.5 Levene's theme

In 1953 the *American Naturalist* published a letter to its editors, of about two pages, by Howard Levene, entitled "Genetic Equilibrium When More than One Ecological Niche is Available." The letter contained a simple model of how a population could remain polymorphic as a consequence of environmental variation in space. Levene's model became a core around which a large theoretical literature developed.[9]

Levene's note began as follows: "In recent years the attention of experimental evolutionists has been directed towards polymorphism as furnishing desirable plasticity to a species" (1953, p. 331). Levene, an associate of Dobzhansky, viewed polymorphism as contributing to population-level adaptive flexibility, and one of his stated aims was to extend the known conditions under which selection would maintain genetic variation in a population. He described circumstances in which polymorphism is retained by individual selection even though the heterozygote is not fitter than the homozygotes in any particular environmental state.

The basic idea is intuitive: perhaps polymorphism can be maintained not through some fixed advantage enjoyed by heterozygotes, but through selection favoring different homozygotes in different parts of the environment. Genotype *AA* might do better in the warm regions, *aa* in cool ones, and polymorphism might be retained though there is no single region in which *Aa* is the best.

There is an interesting conceptual question about this explanation. I presented it as something distinct from heterosis or overdominance, but perhaps it is not so much an alternative as an *explanation* of overdominant fitness relationships. Though *Aa* does not do better in any specific state, it does better on average. After all, no environment is completely homogeneous, and not all individuals with a genotype do equally well, so all attributions of fitness to genotypes must average over some fluctuations, if the model is to have any application to nature. Indeed, it has been argued independently that the fitnesses of heterozygotes and homozygotes used in standard genetic models are best thought of as averages, where a variety of micro-environments are taken into account (Sterelny and Kitcher, 1988; Waters, 1991).

This view is supported by the fact that the simplest way to model the effect of spatial variation in selection does reduce to the basic model of over-dominance. Suppose we determine the average fitness of a genotype by summing its fitness in each different micro-environment, weighted by the proportion of the total environment which is in that micro-environmental state. Polymorphism is then retained if and only if the arithmetic mean fitness of the heterozygote is higher than that of the homozygotes (Dempster, 1955). There is nothing mathematically new here, so perhaps this is not an alternative mechanism to overdominance.

Levene's model did produce something new, as a consequence of a change in the assumptions made about selection. Levene supposes that the environment varies in a discrete way with respect to space, and each individual spends all its life in some particular environmental state, or "niche." Selection occurs locally in each niche, but after the individuals have endured selection, they leave their niches and mate at random in a common pool. The fertilized zygotes disperse at random into the niches. So it is not possible for the population to fragment into subpopulations each genetically adapted to a particular niche; the offspring of a survivor of one niche could equally likely end up in any other niche. The consequences of selection within all the niches in a generation are combined to find the new frequency of alleles in the whole population.

The unusual assumption made by Levene is that though selection takes place within each niche in the standard way, the contribution from each niche to the next generation is fixed and not affected by the pattern of selection in that niche. Levene showed that under these conditions polymorphism will be maintained if the *harmonic* mean fitness of the heterozygote (weighted by the sizes of the niches) is higher than the harmonic mean fitnesses of both homozygotes. The harmonic mean of a set of numbers is the reciprocal of the arithmetic mean of the reciprocals of those numbers. The harmonic mean is always less than the arithmetic mean and less than the geometric mean (unless the numbers averaged are all the same in which case all the means are the same). The harmonic mean is reduced by variance in the numbers averaged, even more than the geometric mean is. Levene's criterion is a sufficient (not necessary) condition for the maintenance of polymorphism. The harmonic mean criterion can be met without the heterozygote being fitter than the two homozygotes in any particular environmental state, or fitter on the (arithmetic) average.

Though this model is mathematically different from the standard model of overdominance, it is still possible to describe it as a case in which the

265

heterozygote is "the fittest," as the heterozygote has the highest harmonic mean fitness. The Levene condition for polymorphism is sometimes described as "harmonic mean overdominance" (Felsenstein, 1976, p. 258). So it would be possible to say that this is another case where population-level complexity is retained as a consequence of the superiority of a certain kind of individual. On this view, a high harmonic mean fitness is as much a property of an individual as any other sort of fitness.

Alternately, the term "harmonic mean overdominance" might be considered a misleading way of describing a situation in which polymorphism is maintained by the fact that AA and aa are favored in different circumstances, and the complex individuals Aa are retained as a byproduct. The model can be viewed in two different ways even though only individual selection is involved. This is not like the situation in section 9.4, where we looked at two different selection mechanisms for explaining heterotic schemes of fitness. In the present case we have one mechanism and two glosses or causal commentaries on it.

Although some may prefer a conventionalist attitude here, there is some basis for preferring the second view. As Levene's theme was developed by others, versions of the model were devised which could not be described in terms of any kind of overdominance – in terms of any kind of advantage to Aa individuals. Overdominance requires that there be three phenotypes. Prout (1968) defines "absolute dominance" as the indistinguishability of AA and Aa with respect to fitness in all niches. Prout described a class of sufficient conditions for retention of polymorphism in a Levene-type population which is broader than Levene's harmonic mean condition, and which includes cases where maintenance of polymorphism is possible with absolute dominance.[10] Let "W_{AAi}" be the fitness of W_{AA} in niche i, and so on. If the W_{AAi} and W_{Aai} terms are equal to each other, the same for all niches, and set to one, then polymorphism is maintained if the *arithmetic* mean of the W_{aai}'s is greater than one but the *harmonic* mean of the W_{aai}'s is less than one.[11] This case cannot be described in terms of overdominance at all. This is the reason for viewing Levene's model as giving an alternative mechanism to the mechanism of polymorphism through heterozygote superiority.

The structural feature of Levene's model responsible for his extension of the conditions for polymorphism was isolated by Dempster (1955). This is the assumption that the terms which describe the size of the contribution made by a niche to the next generation are constants, independent of the pattern of selection in that niche. This is now generally called "soft selection" (Wallace 1968; Christiansen, 1975). The more familiar alternative is for the contribution made by a niche to the next generation to depend on the pattern of selection in that niche – if selection is especially severe in some niche, that niche will make a small contribution to the next generation. This is called "hard selection." This is the assumption implicit in the simpler model of the consequences of spatial variation, which reduces to the standard condition for over-dominance.[12]

The regime described by Levene is one in which polymorphism is retained more readily than it is under a regime of hard selection. If we accept that a polymorphic population is less likely to go extinct than a homogeneous one, under the influence of new changes to the environment, then it is true that the factors which bring it about that selection are soft, are factors which "furnish desirable plasticity to the species," to use Levene's phrase.

However, this is not a regime which, in the usual case, will exist *because* of this plasticity. Soft selection will be found if, for example, a plant produces huge quantities of seeds which saturate every patch of soil (Roughgarden, 1979, p. 232). If there is a fixed number of adults any patch can support, and this number is always reached whatever the genetic composition of the survivors, then selection is soft.

It would be *possible* to develop a group-selectionist claim about soft selection and plasticity, along the lines of Lewontin's 1958 argument about heterosis. It could be argued that populations which (as a consequence of their behavior) tend to experience regimes of hard selection, are more likely to go extinct, as a consequence of their reduced polymorphism, while populations which experience soft selection will survive. But this is an additional claim, subject to the usual controversies about group selection, and not part of the Levene model itself. Similarly, it is widely believed that sexual reproduction does "furnish desirable plasticity" to a species, but it is very controversial whether this is why sex exists. The instrumental properties of polymorphism need not be the same as its teleonomic properties (Chapter 1).

Levene also pointed out in his note that his model describes, in one sense, the "worst possible case" with respect to the retention of genetic diversity (p. 333). This is because new-born individuals have no tendency to settle in niches they are well-adapted to, and there is no tendency for mating to occur within a niche. When expressed this way the Levene model appears analogous to a model of "bet-hedging" by an individual (Chapter 7). At the level of the population, the analog of "tracking" the world would be a correlation between the genotypes of individuals and the micro-environments they inhabit.

There are two ways in which a Levene-type population can track the environment. The first is for individuals to seek out the micro-environment which suits them. Mating occurs in a global pool but dispersal is nonrandom.

This case has been examined by Templeton and Rothman (1981; see also Hedrick, 1986). They found that habitat selection increases the chances for polymorphism under soft selection, and that it extended the conditions for polymorphism under hard selection as well. Templeton and Rothman also model the effects of imposing a fitness cost to individuals for the mechanisms and behaviors required for habitat selection.

The second way in which a Levene population can correlate genotype with micro-environment is for individuals to display more *inertia*. If some individuals mate not in the common pool but locally in their own micro-environment, there will be a correlation between an individual's genotype and its micro-environment. (If all mating takes place locally there is no longer a single breeding population, and the two subpopulations are separate evolutionary entities.) If there is some local mating and some global mating, the conditions for polymorphism are again extended beyond Levene's (Maynard Smith, 1970). There is a continuum between a Levene-type model, in which all or almost all mating is global, and models which examine the effects of small amounts of migration between two populations that mate mostly internally.[13]

So polymorphism can be assisted both through individuals being "smarter" than they were in the original Levene model – seeking out the right micro-environment – and also through their being more sluggish or more parochial in their reproductive proclivities. Both these individual-level properties have the effect of changing the relation between genotypes and micro-environments from a hedging-type relation to a tracking-type relation, and both are favorable to polymorphism.

9.6   The rhythm method

Another literature in which genetic diversity has been related to environmental complexity has focused on a different type of environmental pattern – variation with respect to time.

One of the first papers to look at temporal variation in selection as a mechanism which can maintain polymorphism was the 1955 paper in which Dempster discussed Levene's model.[14] Dempster recognized the basic characteristic that genetic models of temporal variation must have: the property of a genotype that determines its fate is its geometric mean fitness over time. A large literature has discussed models constructed along these lines.

Some basic results are as follows. With haploid genetics no polymorphism is possible unless the geometric mean fitnesses are exactly equal (Dempster, 1955). In a diploid system if there are three phenotypes then polymorphism is maintained if the geometric mean fitness of the heterozygote is higher than the geometric mean fitnesses of both homozygotes (Haldane and Jayakar, 1963; Gillespie, 1973).[15] With absolute dominance polymorphism can be maintained if the arithmetic mean fitness of *aa* is greater than that of *AA/Aa* but the geometric mean fitness of *aa* is lower (Haldane and Jayakar, 1963). Note that as polymorphism is possible with dominance but not with haploids – two cases which are similar at the level of phenotypes – this is an example of an underlying genetic property making a qualitative difference to a model of selection.

Some writers explicitly stress the role played by variance in fitness in models of temporal fluctuation in selection (Gillespie and Langley, 1974). If a heterozygote has less variance in fitness than the homozygotes, then even if it is not fitter than the homozygotes at any specific time, polymorphism can be maintained, and this is sometimes referred to as "geometric mean overdominance." This in turn is related by some writers to the possibility, discussed earlier, that heterozygotes are more homeostatic than homozygotes (Hedrick et al., 1976, p. 13).[16]

So some discussions explicitly stress the continuities between these newer models and the older ideas about homeostasis and buffering. As I have argued, there is a basic pattern of explanation that all these models share: they explain organic complexity in terms of environmental complexity. However, it would be a mistake to see newer models of fluctuating selection as simply describing the underlying basis for the older ideas. For example, although we might say that a genotype with low variance in fitness has a property that resembles homeostasis, this need not involve any property of organic stasis maintained by individuals with this genotype. Low variance in a genotype's fitness *might* be produced by individual-level homeostasis, but it does not have to be. If heterozygotes are intermediate with respect to some phenotypic property, for example, this could produce low variance in fitness across individuals, without any "buffering" or extra homeostasis in the physiology or development of those individuals. In the models of Levene, Haldane and Jayakar, Gillespie and others, the heterozygote *genotype* can be favored without it being the case that all heterozygote *individuals* have higher fitness than homozygotes.

In this chapter we have seen, firstly, that if heterozygote individuals are more homeostatic than homozygote individuals, polymorphism can be the result. We have also seen that if the heterozygote genotype has reduced variance in fitness, polymorphism can result, even if the individual heterozygotes are not especially homeostatic.

A genotype cannot be homeostatic in the narrow sense outlined in Chapter 3. That definition applies only to individual organisms. But let us say a genotype can be "quasi-homeostatic." A genotype is quasi-homeostatic if some property of stasis or reduced variation associated with the genotype tends to make the genotype successful, where this property of stasis is the product of some other property of complexity or variability associated with the genotype. So quasi-homeostasis, like genuine homeostasis, involves both variability and stasis, and it is again required that the benefits associated with variability go via a benefit associated with stasis in some other organic property.

Some models of the consequences of variation in selection, such as Gillespie's and maybe Levene's, can be viewed as explaining polymorphism in terms of the reduced variability of the heterozygote genotype. But these are not models of "quasi-homeostasis" in the sense just defined. Stasis plays a special role, but this stasis is not the product of some other property of variability. Later in this chapter we will look at a genuine case of quasi-homeostasis.

We should also note that in the case of the Levene model and the case of temporal variation in selection, there can be polymorphisms maintained that cannot be described in terms of overdominance or heterozygote superiority of any kind, as they involve absolute dominance. With absolute dominance there are only two phenotypes. So models of temporal and spatial variation in selection cannot be understood in terms of a simple extension of the idea of heterozygote advantage, or of homeostasis as a special organic property.

I will not go into more recent and elaborate developments of these ideas here. Models of environmental variation continue to occupy an important place in the (still controversial) project of explaining polymorphism in terms of forces of selection. The program has been pursued with particular vigor by J. Gillespie (1978, 1991).[17] All the models discussed in the last few sections are c-externalist. Within this basic framework, the emphasis has shifted over time from models in which heterozygote individuals have special properties, towards models in which different genotypes are favored in different circumstances.

In the remainder of this chapter I will try to tie some parts of this discussion of population-level properties to the earlier discussions of individual-level flexibility. One goal of these last sections is finding principles that might link individual-level and population-level complexity.[18] We will start this final task of the book by looking in detail at a particular work, well-known but controversial, which purports to give a general treatment of the evolutionary consequences of environmental variability. This is Richard Levins' book *Evolution in Changing Environments* (1968a).

Levins' work is important here first because it does seek to give a general account of relations between individual-level and population-level complexity. This account will be evaluated in its own right. Second, Levins' book has interest because of its unique historical location. Levins' work was one of the first to make extensive use of optimality arguments in place of orthodox genetic models. But Levins' book is also a contribution to a program discussed in section 9.4 above. This is a line of thought which treats population-level adaptation as similar in status to individual-level adaptation, and in particular recognizes polymorphism as a population-level "response" to environmental complexity. This line of work includes some explicitly group-selectionist claims (Lewontin, 1957). Although Levins' work does not look like a product of group-selectionism, I will argue that one feature of his basic model should be viewed as embodying this type of group-selectionist thinking. If Levins' model is to be used within a theory that assumes only individual selection, then a change must be made to its structure.

9.7 Levins' machinery

Levins aims to view a range of features of individuals and populations as adaptations to patterns of environmental variability in space and time.[19] Suppose, he says, that large size is an adaptation to cold, as it helps the animal retain heat. That type of explanation is well-understood. But what should we expect in an environment which is sometimes hot and sometimes cold? There are four possibilities: (i) an intermediate size, as suggested by "folk liberalism," (ii) adaptation to one condition only, at the expense of fitness in the other, (iii) a polymorphism in the population, with some individuals adapted to each condition, and (iv) tracking of the environment; the animal tries to be large in the cold and small in the heat (1968a, p. 10).

Some of these options we have examined already. Chapter 7 aimed to give some conditions for when individuals should track the environment rather than adapt just to the most important single state. We have not looked at the possibility of intermediate "cover-all" phenotypes which are not optimal in any particular state. Levins' third option – polymorphism at the population level – we have examined with genetic models. But no attempt has been made here to compare all of them at once. This is what Levins seeks to do.

Suppose that, in principle, both the environment and the individual's phenotype can vary in a continuous way. For every combination of phenotype and environment there is an associated fitness value. One pattern this relationship could have is expressed in Figure 9.1. There is a bell-shaped ridge of high fitness running diagonally along the graph. For any environmental condition there is one best phenotype, and fitness falls away symmetrically on either side of this peak.

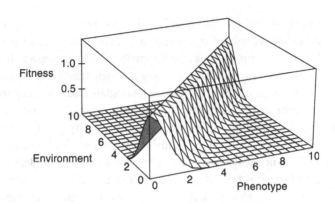

Figure 9.1. Fitness as a function of phenotype and environment

An example used by Roughgarden (1979) is ideal. Think of both the environmental variable and also the phenotypic variable as color, in some single dimension. High fitness results from camouflage, from a match between phenotype and environment. Some camouflage is bestowed by an imperfect match, and the benefit falls off on either side of the optimum.

If the organism is not able to actively select the right patch or change its color individually, how will this environmental variability be dealt with?

If we look at Figure 9.1 from the left hand side, with the environment axis in front, we find that for any phenotype there is a single best environment, and a bell-shaped curve describes how well the phenotype does in other conditions. If the curve is very spread out, each phenotype does well in a large range of environments. If the curve is sharply peaked it is only good in the ideal condition. The spread of these curves can be regarded, Levins says, as a measure of the "tolerance" of that phenotype for non-optimal environments. He also calls this tolerance a measure of homeostasis (1968a, p. 14). Levins makes the assumption that the areas under all the curves are the same. We assume they have the same shape as well.[20]

Second, we can look at the diagram from the right-front view, and view phenotype as a continuous variable. For any environment there is a best phenotype, a curve describes how sharply fitness drops when an inappropriate phenotype is produced.

Now suppose the only two environmental conditions actually encountered are those where the environmental variable has value 3 or value 6. Figure 9.2 is a projection of that part of Figure 9.1 into two dimensions, viewed from the right-front perspective.

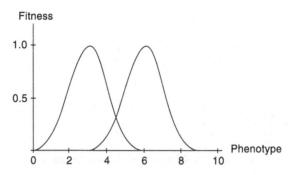

Figure 9.2. Fitness as a function of phenotype, for environmental state 3 on the left and 6 on the right.

These curves are sufficiently far apart that when they are averaged, the average of the curves has a minimum in the center and two peaks near the original peaks. If the curves were close enough together for their inflection points to overlap, the average curve would have one peak around the center of the graph. For example, if the envionmental states encountered were 3 and 4, rather than 3 and 6, the inflection points would overlap.[21]

Levins uses optimization methods to determine the consequences of these different types of case. First we represent the information in Figure 9.2 differently by having the horizontal axis represent fitness in one environment (W_i) and the vertical axis represent fitness in the other (W_j). Figure 9.3a represents the case when $i = 3$ and $j = 4$, and Figure 9.3b represents the case when $i = 3$ and $j = 6$. In the former case the outside of the set is convex in shape, and in the latter case it is concave.[22] Thus we can distinguish between what Levins calls *convex* and *concave fitness sets*.

These curves represent the biologically possible combinations of payoffs in a situation. A point on a curve represents the combination of payoffs for a particular phenotype.

Levins seeks to predict what the population will do in these different cases with the aid of optimization principles; it is assumed that evolution will tend to optimize some property. The properties Levins uses are measures of mean fitness.[23]

Levins uses two different optimization criteria, each associated with a specific type of environmental patterning. He calls them "fine" and "coarse" environmental "grains." One criterion ("fine") uses an arithmetic mean, and

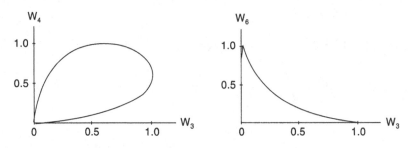

Figure 9.3a and b. Convex and concave fitness sets.

the other ("coarse") uses a geometric mean. Grain is fine, according to Levins, if the individual encounters a sequence of (possibly) different states in its lifetime, and deals with the average. Grain is coarse if the individual tends to spend its whole life in one environmental condition or the other.

Levins assumes that when variation is fine grained, evolution will bring it about that the population will display the phenotype or combination of phenotypes associated with the highest arithmetic mean fitness, as measured by L_A.

(22) $\quad L_A = P_i W_i + P_j W_j$

The terms P_i and P_j here represent, strictly speaking, the *proportions* of the environment which are in S_i and S_j. In some cases these can be interpreted as probabilities – the probability of an organism finding itself in an S_i patch, or the probability of next year being an S_i year. But in other applications it may be better to speak simply of proportions of the environment, and to say that L_A determines mean fitness by weighting fitness by the proportion of the environment in each state.

Any particular value of L_A is associated with a variety of possible values for W_i and W_j. The combinations of W_i and W_j associated with a given arithmetic mean fitness lie on a straight line on a graph of the type in Figure 9.3. Levins calls these lines "adaptive functions," but I will just call them lines of equal average fitness, or average fitness lines. The slopes of the lines represent different values for P_i. The further away from the origin (0,0) the line is, the higher the average fitness.

Not all combinations of W_i and W_j are biologically possible – only those that lie on the fitness set. Levins claims that the point on the fitness set associated with the highest mean fitness will be the phenotype that is found. Finding the point in the fitness set which touches the average fitness line

Complex individuals, complex populations

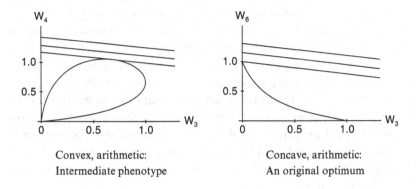

Convex, arithmetic:
Intermediate phenotype

Concave, arithmetic:
An original optimum

Figure 9.4a and b

furthest from the origin, is finding the point on the fitness set associated with the highest arithmetic mean fitness.

When the fitness set is convex and optimization is arithmetic, an intermediate phenotype is best; a cover-all which is distinct from both the original optima. Convexity generates the moderate middle course, or "folk liberalism." When the fitness set is concave the best phenotype is well adapted to one environmental state only. On a concave fitness set the jack of all trades is truly master of none, and not master of the average trade either. These cases are illustrated in Figure 9.4. Here P_3 is assumed to be 0.2. With the convex fitness set the best phenotype has a value of about 3.85. With the concave fitness set the best phenotype is about 6.[24]

Levins claims that when the grain of environmental variation is "coarse" rather than "fine," evolution will maximize geometric mean fitness, measured as follows:

(23) $\quad L_G = W_i^{P_i} W_j^{P_j}$

As before, the aim is to find the point on the fitness set associated with the highest L_G. Different combinations of W_i and W_j which generate the same geometric mean fitness can now be represented as hyperbolas on the graphs.[25] The furthest-out hyperbola represents the highest attainable geometric mean fitness, and the point where it touches the fitness set determines the best phenotype. The shape of the hyperbolas depends also on the value of P_i.

If the fitness set is convex, things are much the same as before. The best hyperbola touches at an intermediate "folk liberal" phenotype. If the fitness set is concave the situation is more complicated.

275

Consider first the idea of a *mixture* of phenotypes C_i and C_j. This mixture has fitnesses in both environmental states. A mixture with mostly C_i and a few C_j will do better when the environment is in state i – it will have a high W_i but a low W_j. The mixture's fitness in an environment is the arithmetic mean of the fitnesses, in that environment, of the phenotypes comprising the mixture. Each mixture will have a combination of fitnesses representable with a point on the same graphs as before. The fitness combinations for different mixtures of two phenotypes are represented on straight lines connecting the points for those two phenotypes. A mixture that is almost all C_i will be close to the point for C_i itself, a 50/50 mixture of C_i and C_j will be half-way between the two, and so on. We are dealing here with a single mixture produced on every trial (see formula (13) in Chapter 7 above).

If the fitness set is convex, there is no way a point representing a mixture can touch an average fitness line further from the origin than a line which touches the curve itself. If the fitness set is concave, the line representing mixtures of the two original optima lies outside the fitness set. A mixture of the two original optima is more fit on this measure than any single phenotype.

In Figure 9.5, again P_3 is assumed to be 0.2. In the convex case the optimum is a phenotype of 3.80. In the concave case the optimum is a mixture of phenotypic values of 3 and 6, where 3 is produced in proportion 0.19.

These are the results of Levins' basic model. Three different options – a cover-all intermediate phenotype, adaptation to a single condition, and polymorphism – are predicted for different cases.

There are also various ways to introduce tracking of the environment into this framework. One way is to suppose that the phenotype expressed by an

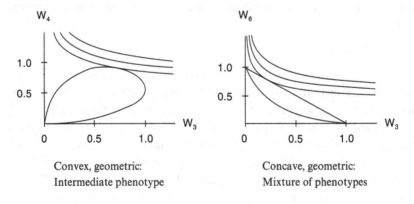

Convex, geometric:　　　　　　　Concave, geometric:
Intermediate phenotype　　　　　Mixture of phenotypes

Figure 9.5a and b.

individual is dependent to some extent on the environment. Suppose that when the environment is in S_i, the phenotype produced is displaced towards the optimum for S_i, and when S_j occurs the phenotype is displaced towards the optimum for S_j. Here the adaptive modification of phenotype by environment is only partial, but there is perfect tracking. Levins shows that the consequence of this is that the fitness set is made more convex (see especially Levins, 1968b). A single general-purpose type of individual (now a flexible type) is more likely to be the optimal solution.

Another way to model tracking is to suppose that individuals have a cue which provides imperfect information about the local state of their environment. We can then represent posterior probabilities, the probabilities of environmental states conditional upon the state of the cue, in the average fitness lines on the graphs. There may be separate optimal phenotypes for each state of the cue, each derived in the way outlined above. An optimal individual may then develop as a function of state of the cue (see Levins, 1963). Again, the result is an extension of the conditions in which a single type of individual, rather than a polymorphism, is the optimal solution.[26]

9.8   The coarse and the fine

The next two sections will look at two distinct problems with this framework. The first, which is minor, concerns the concepts of fine and coarse environmental grain. The second has to do with the relations between individuals and populations.

We have several times encountered different ways of averaging and tallying payoffs. In Chapter 7 when treating individual plasticity, several different types of averaging were distinguished. The genetic models in the present chapter also made use of a variety of measures. The consequences of spatial variation with hard selection are determined by the arithmetic means of genotype fitnesses; the case of spatial variation with soft selection (Levene model) is controlled by the harmonic means of genotype fitnesses, and the case of temporal variation is controlled by the geometric means of genotype fitnesses.

If we disregard the Levene case, then in genetic models there is a link between spatial variation and arithmetic means on the one hand, and temporal variation and geometric means, on the other.

In a series of papers Levins wrote which preceded the 1968 book (especially Levins, 1962), he used the difference between space and time to differentiate the arithmetic and geometric cases. In the 1968 book he says arithmetic means are used for "fine" and geometric means for "coarse" grains of environmental variation. Whether variation is coarse or fine is determined by the relations between the environment's intrinsic patterns and the organism's life-history properties and behavior. An organism which never moves will encounter a spatially variable environment differently from a mobile one. The nonmobile organism will find itself in some specific condition and spend its whole life in it, while a freely moving organism will encounter a sequence of states. Levins' move to the concepts of "coarse" and "fine" grain is a way to take account of this role played by the organism in determining the pattern of its experience.

However, there are problems with the specific way Levins' concepts achieve this. The determinant of which measure is appropriate is the relations between the stakes and payoffs of different trials. Whether payoff events are related arithmetically or multiplicatively is not determined by *either* the distinction between spatial and temporal variation *or* the distinction between fine and coarse in Levins' sense. For example, if the environment switches rapidly between different states in time, and the organism encounters many different states, this counts as a fine grained variation for Levins. But this could be a case in which either arithmetic or geometric means are relevant. If the organism experiences successive events in a way which is multiplicative, the geometric mean is the relevant measure. If its payoff from an event is proportional to the present "balance" of fitness, and a zero payoff cannot be recovered from, then a geometric mean is needed. In genetic models the biological relations between successive generations are structured such that temporal variation across generations makes geometric mean properties the relevant ones; reproduction across generations is multiplicative. But the relation between different trials or events within an individual's life can also be multiplicative. For example, if an individual has to survive several distinct periods in its life, and there are different survival probabilities associated with each period, these survival probabilities are multiplied to get the probability of surviving through all the periods.

In the abstract there is no rigid connection between *either* the space/time distinction *or* the many-states/one-state encountered distinction, on the one hand, and the distinction between payoff structures that involve geometric means and those that involve arithmetic means, on the other.

9.9 A counter-example

In this section we will look at a case which reveals a problem in Levins' model that has to do with the relations between individual-level and population-level properties of complexity. This case will be used later to draw some general conclusions about complexity properties.

Levins says that when the relevant measure of mean fitness is geometric, and the fitness set is concave, we should expect polymorphism (Figure 9.5b). We should find a mixture of phenotypes in the population, each well-adapted to some specific environmental state.

Let us first understand this polymorphism in the standard genetic sense. The prediction is of a population of individuals with different genes, some individuals adapted to S_i and some to S_j, with the exact composition of the population determined by the probability of S_i. A simple way for this to be realized is by a population of asexual, haploid individuals whose phenotype is determined by one locus. Let us now look at a case adapted from one given in Seger and Brockmann (1987).[27]

The environment is variable in time. Each individual lives for a season. This is "coarse grained" variation, by Levins' definition. Wet seasons and dry seasons occur with equal probability and as independent trials. Suppose there are three genotypes I_1, I_2 and I_3. I_1 is a wet-year specialist and I_2 is a dry year specialist. These genotypes produce the same individual phenotype in all cases. I_3 is a genotype for "developmental roulette," or coin-flipping. It produces the phenotype associated with I_1 with probability 0.25 and the phenotype associated with I_2 with probability 0.75. So in any one year, I_3 will be carried by some individuals with the wet-year phenotype and some individuals with the dry-year phenotype.

The genotypes' fitness properties are represented in Table 9.1. In a wet year, I_3 has a fitness of 0.7 because it produces a 25/75 mixture of phenotypes with fitnesses of 1 and 0.6.

Suppose first that I_1 and I_2 are competing in the population. I_2 has a slightly higher geometric mean fitness than I_1. We assume that reproduction is asexual, replication rate is simply proportional to fitness, and the population is very large. Then if there is a long and representative series of years, I_2 will

Table 9.1 Fitness properties of three genotypes

	I_1	I_2	I_3
Wet Year	1	0.6	0.7
Dry Year	0.5	1	0.875
Geom. Mean	0.707	0.775	0.783

eventually win out over I_1 and become fixed, owing to its higher geometric mean fitness.[28] This occurs even though the fixation of I_2 does not result in the maximization of population geometric mean fitness, as measured by L_G. That measure is maximized when the population is polymorphic, with the frequency of I_2 at 0.75. If we do a Levins-style analysis of this case the prediction is this polymorphism. But in fact in this scenario I_2 will out-compete I_1.

Now suppose I_3 appears. This genotype produces individuals which embody the complexity that would be realized by the polymorphic *population* with the highest value of L_G. Genotype I_3 has a geometric mean fitness equal to the population geometric mean fitness of the optimal polymorphism of I_1 and I_2. As I_3 has a higher individual geometric mean fitness than either I_2 or I_1, it will become fixed. Once this is complete, we will have a population which looks like one with a genetic polymorphism, but which is not genetically polymorphic. It is a simple population of complex individuals. This is a case where complexity embodied in the individual will out-compete simple individuals which could realize, as a population, the same pattern of complexity.

Complexity is favored here without any individual tracking of the environment by I_3 individuals. Allele I_3 is bet-hedging, in the sense of Chapter 7. It produces the same mixture of phenotypes every year. If I_3 individuals could receive some information about what kind of year they are in, and alter their developmental trajectory appropriately, then they could do better again.

We should note that the "individual complexity" associated with I_3 is of a particular kind. It is not a case in which a complex individual does a variety of things; the individual is complex with respect to its *capacities*. A single type of individual can produce a variety of phenotypes, although each token individual produces only one phenotype.

This case is a problem for Levins' analysis because his framework does not distinguish between genetic polymorphisms and "pseudo-polymorphisms" which arise from complexity in the individual. Levins' model predicts correctly that I_3 will win out over I_2. But it also predicts incorrectly that if I_1 and I_2 exist in a population alone, what should be found is the polymorphism which maximizes L_G. In fact I_1 and I_2 will not coexist in a polymorphism maintained by individual selection.

This case can be used to do two different things. In the next section it will be used to further investigate the explanatory structure of Levins' model. Later it will be used to discuss in a more general way the relations between individual-level and population-level complexity.

9.10   The group-selectionist structure of Levins' model

Levins' model encounters a problem with the case described above because it treats two different kinds of variability in a population as equivalent. It does not distinguish between variability that is the consequence of an individual capacity for variable behavior, and variability that is the consequence of a population-level property of polymorphism. Given this fact, we must make some change to the model before it can function as an appropriate framework for the investigation of the consequences of environmental variability.

There are two different changes (at least) which can be made at this point. One change takes the form of a restriction of the domain of the model. The other takes the form of a reinterpretation of the model itself.

The first possible modification I will call the "individualist" view. We can restrict the application of the model so that all variability in behavior addressed by the model is understood as variability in the capacities of individuals. The model describes conditions for pseudo-polymorphism rather than polymorphism. The model would then provide a framework for assessing optimal strategies for individuals or genotypes in the face of environmental complexity. It would not say anything about complexity as a population-level property, except that population-level complexity can exist as a byproduct of complexity in individuals' capacities. The measures of mean fitness used in the analysis would be understood as properties of individuals or genotypes, the usual fuel for evolutionary change.

As was stressed in section 7.9, the individual "mixing" of phenotype or behavior must be interpreted here as a within-trial property. Switching phenotype from trial to trial will not do an individual any good. Only mixing of phenotype by the individual within a trial, or mixing by a genotype of the phenotypes of individuals with that genotype, is beneficial.

If the model is viewed in this restricted way the problem raised in the previous section does not arise. The model as reinterpreted predicts, correctly, that in a representative sequence of years I_2 will do better than I_1, and I_3 will do better than either I_1 or I_2. The only problem with this reinterpretation is that it gives up on a distinctive goal that Levins' discussion has – it gives up on the attempt to treat individual-level and population-level complexity properties simultaneously within a single framework.

There is also another way to address the problem raised in the previous section, which does not give up on the attempt to treat individual-level and population-level complexity in a unified way. This option is to view Levins' model as making an appeal to group selection.

Models

Before outlining how this view of the model works it is necessary to briefly discuss a general issue about fitness properties. Levins makes use of two properties of mean fitness, L_A and L_G. There are two different ways in which mean fitness properties can be used in models of this sort. They can firstly be used simply to plot or predict a trend, without necessarily giving causal information about the mechanism producing the trend. Alternately, they can be used to describe part of the causal mechanism. Properties of population mean fitness usually have the former status. For example, in the simplest one-locus population genetics models it is possible to predict what will happen to a population by attending to a measure of population mean fitness ("$\overline{W}$"). It can be shown that in certain circumstances this measure will be maximized.[29] Similarly, it had been shown that evolution in Levene's model can be described in terms of maximization of a measure of geometric mean fitness – an analog of L_G (Li, 1955; Roughgarden, 1979; see also Levins and Macarthur, 1966). But these properties are abstractions which describe the population as a whole; they are not fitness properties of the type that drive evolutionary change. These processes are driven by differences in individual fitness, as represented in the fitnesses of genotypes.

In most models, properties of population mean fitness are, at best, useful abstractions which can be used to plot or predict the course of evolution without being causally salient in the process. But there is an exception. If a process of group selection is occurring, then in some cases a property of population mean fitness might be a good measure of that population's likely success in competition with other populations. Populations with high mean fitness might be the ones which grow, spawn new colonies and tend not to go extinct. In that case, population mean fitness is analogous to the fitness of a genotype in the standard models; population mean fitness is part of the causal engine producing evolutionary change.

The population mean fitness properties in Levins' model are usually viewed as heuristics or predictive tools. (See, for example, Roughgarden, 1979.) Indeed, if individual selection is supposed to be the mechanism, that is the *only* way they can be viewed. The alternative is to view the measures of mean fitness in Levins' model as causally salient in processes of group selection. Properties of population mean fitness are then taken to be measures of the likely evolutionary success of a population when compared to others. If populations with high mean fitness are fit groups, and will do well against other groups for that reason, then a population polymorphic for I_1 and I_2 can be favored in a process of group selection over a population of all I_2 individuals.

282

If that is how the model is interpreted, then the two different ways of bringing about the same population geometric mean fitness – monomorphism for I_3, and polymorphism of I_1 and I_2 – can be regarded as in one sense equivalent. The property we take to be causally important is the fitness of the population, as measured by L_G, not the details of how this population fitness is brought about. Polymorphisms and pseudo-polymorphisms can be seen as ways to realize a population-level property which is viewed as causally important.

This group-selectionist interpretation of Levins' 1968 model is not something I have invented to deal with the problems discussed here. This is the *explicit* framework in the series of papers Levins published before the 1968 book. In the first paper of the series he said: "we will determine which population characteristics would provide the maximum fitness, where fitness is defined in such a way that *interpopulation selection* would be expected to change a species toward the optimum (maximum fitness) structure" (1962, p. 361, italics added. See also Levins, 1963, p. 83). This is treating population fitness not as a predictive abstraction, but as fuel in a higher-level process of selection. This explicit appeal to group selection does not appear at all in Levins' 1968 book. My claim is that it has left its mark in the structure of the model itself.

On this interpretation, Levins' work is the culmination of a line of thought discussed earlier in this chapter, which treated individual-level and population-level adaptation in a unified way. Considered as an explanation of polymorphism, Levins' model under the group-selectionist interpretation has the same explanatory structure as the view of heterosis proposed in Lewontin (1958). Polymorphism is a group-level adaptation to environmental complexity.

The *impact* of Levins' book is another matter. The book had impact because it introduced methods which were valuable within the emerging program of individualist optimality analysis in biology.[30]

Viewing Levins' model in this way makes it comprehensible why polymorphisms and pseudo-polymorphisms are not distinguished. This is not to say that the model is then a complete group-selectionist account of complexity.[31] It has the familiar problem faced by many group-selectionist models: can the group-selection process succeed in the face of individual-selection in the other direction? For though a polymorphism of I_1 and I_2, and a monomorphism of I_3 give rise to the same group fitness on measure L_G, and may have similar group-selective advantages, the polymorphism of I_1 and I_2 is not stable under individual selection. Within any group I_2 will always tend to be increasing in frequency. On the group-selectionist construal I_1 in some respects plays the role of an allele for altruism. It is not favored within any group, but its presence raises the fitness of the group. (See also Cooper and Kaplan, 1982, for a related argument.) A difference is that in models of altruism group fitness is usually an always-increasing function of the frequency of altruists, while in the Levins model group fitness (as measured by L_G) is highest at an intermediate frequency of I_1. Another difference is that an altruism gene usually raises group fitness *by* raising the fitness of the other individuals. But there is a partial analogy here, as I_1's presence does serve "the good of the group," in a certain sense. So this view of Levins' work links his models up with another whole family of problems in evolution.

9.11   Quasi-homeostasis

By making use of group selection we can understand some otherwise puzzling features of Levins' model. But we also make the model more controversial and problematic in other ways. If our investigation of complexity is to make use of individual selection only, then as outlined earlier, the only way to make use of Levins' model is to view it in a restricted, individualist way. It does not give a suitable general account of the relations between individual-level and population-level complexity.

Levins' model does not enable us to integrate polymorphism with individual-level plasticity into a single theory of the consequences of environmental variability, if individual selection is to be the only mechanism allowed. But the preceding discussion does tell us something about the relation between the two kinds of complexity: when an organic system encounters environmental complexity in a way which makes measures of geometric mean fitness relevant, then complexity realized in individuals can be favored over a polymorphic condition. Complexity realized in individuals' capacities can reduce variation in payoff across trials.

The circumstances in which this advantage exists are restricted, but there is a range of possible cases where it can apply. We have looked at the advantage associated with a genotype which varies its expression across individuals. This is an instance of what was labelled "quasi-homeostasis" in section 9.6. A genotype is quasi-homeostatic if some property of stasis or reduced variation associated with the genotype tends to make the genotype successful, where this property of stasis is the product of some other property of complexity or variability associated with the genotype. This is what we see in the case of a genotype like I_3. I_3 does well in competition with I_1 and I_2, even though it only produces phenotypes which these other genotypes can produce. Its superiority is a consequence of its reduced variance in fitness across generations. This reduced variance in fitness is, in turn, the consequence of the variation in phenotype that I_3 produces. A property of variation or heterogeneity gives rise to a distinct property of stasis, and this property of stasis produces success in the long term.

We should also note that in the simple scenario assumed in the model of section 9.9, the long-term prediction is the same for genotype I_3 which mixes its expression between individuals, and a genotype I_4 which produces individuals that each mix or switch between two phenotypes, in the optimal proportion. The predictions are the same as long as the switching of behaviors is cost-free, the population is large, and I_4 individuals *each* gain the same arithmetic mean payoff within a given year that I_3 does *across* its individuals in that year. The geometric mean fitness of genotypes I_3 and I_4 in a long representative sequence of years will be the same. So I_4 is a quasi-homeostatic genotype as well.

In the case of I_4, each of the individuals with the genotype engages in complex behavior. But the I_4 *individuals* do not benefit from this. An I_4 individual does worse in a wet year than an I_1 individual, and does worse in a dry year than an I_2 individual. The expected fitness of an I_4 individual, within a year, is less than that of an I_2 individual; the I_4 individual's variability does not make it an adaptively homeostatic individual. An I_4 individual is in fact doing something quite strange; it is varying its behavior randomly even though the world is in a single state throughout its life. But again, in a long representative sequence of years, genotype I_4 will win out over I_2.

The result involving arithmetic and geometric means does not only apply to genotypes. An individual organism can also obtain this benefit in some circumstances. An individual which produces simultaneously two forms of an enzyme when it must deal with two environmental conditions that vary in time engages in the same type of mixing within trials. An individual which does a variety of things in sequence during the course of a single "trial" can also obtain this benefit, as long as accounting is arithmetic within a trial and geometric between them. And as we have seen, the benefit can also accrue to a group.

This is not an instant recipe for biological complexity. It is only operative in a restricted set of circumstances, and we are concerned with a very low-level type of organic complexity. But this factor does appear in a variety of evolutionary models, having similar consequences in many cases.

9.12 Summary of Part II

The primary aim of this second part of the book has been to outline and compare a range of simple models investigating the value of flexible responses to environmental heterogeneity. Three mathematical frameworks deriving from different fields share the same basic structure. The biological model of plasticity, the Bayesian model of experimentation, and signal detection theory all focus on a common set of relationships between heterogeneity, probability, cost and benefit. The biological model of Chapter 7 describes the reliability properties that an environmental cue must have before it can be a useful guide to behavior. The Bayesian model of section 7.6 measures the value of an experiment used in decision-making. This "value" is understood in a subjective sense in the Bayesian model, as are the probabilities, while the biological model makes use of an objective, biological sense of payoff and a physical interpretation of probability.

In Chapter 8 the biological model of plasticity was linked to the framework of signal detection theory. Signal detection theory is concerned with specifying the *optimal* cue for an agent to use in the control of behavior; it describes the best way to make use of the resources provided by environmental signals. In addition, the signal detection model was used to address the question of the special costs that may be associated with flexible strategies. Signal detection theory provides a sophisticated way of asking questions about the practical value of reliability properties. One moral that emerges from this approach is the two-dimensional nature of reliability.

The framework outlined in these two chapters also builds on and sharpens the more informal specification, in Part I, of the environmental conditions in which cognition is valuable. Cognition makes possible functional complexity with respect to behavior, and this complexity tends to be valuable in environments containing a certain combination of variability and stability: variability in distal conditions relevant to the agent's well-being, and stability in the links between these distal conditions and more proximal and observable conditions. That much was stated in Part I. In Chapters 7 and 8 we have looked at some of the mechanics of these relationships; we have seen how different types of environmental heterogeneity interact with cost/benefit asymmetries and with reliability properties, to determine the value of various kinds of functional complexity.

In this final chapter of the book a different set of problems has been introduced. What is the relation between individual-level functional complexity and population-level properties of complexity? Is the same set of concepts involving environmental heterogeneity, reliability, cost and benefit important here as well? These problems are so complicated that my treatment of them has been cautious and exploratory. Genetic polymorphism at a single locus was used as an example of a property of population-level complexity, and we looked at an intricate line of research linking polymorphism to properties of environmental complexity. This is one example of a c-externalist approach to population-level properties.

Lastly, we looked at one large scale synthetic treatment of individual-level and population-level complexity – the central model in Richard Levins' *Evolution in Changing Environments*. I argued that the treatment of polymorphism in Levins' model can work only if the model is understood in terms of a mechanism of group selection, and also discussed how geometric mean fitness measures can favor individual-level complexity over polymorphism in certain cases. But I have only scratched the surface of this last set of issues.

Notes

1 There is also a range of other ways in which relations between individual-level and higher-level complexity can be addressed. Some would furnish good case studies here. For example, J. T. Bonner (1988) claims that there is an inverse relation in any given environment between the individual complexity of organisms, measured as diversity of cell types, and the number of species at that level of complexity. For Bonner this relationship is in turn connected to the positive correlation between individual complexity and size. Bonner's treatment of complexity exhibits an interesting mixture of externalist and internalist arguments. See also Gordon (1991) for a discussion of different types of organic heterogeneity which distinguishes sharply between variability existing among individuals at a time, and change by individuals across time.

2 See Roughgarden (1979) for a good introduction to population genetic models, with a specific focus on some of the issues we are concerned with.

3 A mutation/selection balance is another simple mechanism which will maintain polymorphism, but here I am concerned just with purely selective mechanisms. Genetic drift will also be ignored through most of this chapter.

4 If *Aa* is the *least* fit of the three genotypes there is also an equilibrium frequency at which the population will stay. But this is an unstable equilibrium. A slight movement in either direction is followed by more movement in that direction, until one allele is lost.

5 For a discussion of the sickle cell case, see Cavalli-Sforza and Bodmer (1971).
  As Dobzhansky's work was focused on heterozygotes for whole chromosomes, or whole stretches of chromosome, rather than single loci, a variety of other

explanations for heterozygote superiority are possible here also. For example, Dobzhansky (1951) suggested that as there is no crossing-over in the chromosome inversion heterozygotes he studied in *Drosophila*, heterozygosity will be favored as it prevents the breaking up of co-adapted gene complexes (pp. 123–4). This is not an explanation that has to do with adaptive variability. It has to do with adaptive stasis, in fact – the preservation of a specific, adaptive organic structure through the uncertainties of sexual reproduction.

6 See also Thoday (1953) for an early and cautious presentation of this idea. Another cautious, but more recent, endorsement is Roughgarden (1979, p. 37).

7 See also Waddington (1957b) for a plea for more precision in discussions of genetics and homeostasis.

8 For reviews of models of group selection, see Wade (1978), and Wilson (1983). For philosophical discussions, see Sober (1984), (1993) and Lloyd (1988).

9 A clear and accessible review is found in Roughgarden (1979). More detail, without a great deal of mathematics, can be found in Hedrick et al. (1976) and Hedrick (1986). Felsenstein (1976) gives a denser but also conceptually acute review. For a more mathemetical treatment see Christiansen and Feldman (1975).

The literature surrounding the Levene model is not the only literature giving explanations for genetic polymorphism in terms of environmental variation in space. There is also a large literature on "clines," in which patterns of genetic variation which exist *in space* are related to spatial patterns in the environment. Again, see Roughgarden (1979) for an accessible review and Felsenstein (1976) for a more compressed and detailed one. The large literature on "allopatric" speciation – speciation caused by a spatial barrier to gene flow – provides a different type of example. See also Ford (1975) for detailed discussion of the relations between ecological and genetic properties.

10 The term "protected polymorphism" is used for cases in which an equilibrium frequency might not be known, or might not exist, but it is known that no allele can be completely lost from the population through selection (Prout, 1968).

11 A haploid population can face the Levene situation, with local soft selection and no tendency for offspring to inhabit their parent's niche. It turns out, in an unusual result, that the conditions for protected polymorphism in a haploid case are the same as those with absolute dominance (see Hedrick et al., 1976, p. 17). In contrast, there is no way for spatial variation with hard selection and global mating to generate a protected polymorphism with absolute dominance or haploid genetics. There is also no way for selection to retain polymorphism with variation in time, given haploid genetics – see section 9.6.

12 Levene's model can be converted to a model of hard selection by changing the definition of the weights associated with each niche (Christiansen, 1975).

13 See Roughgarden (1979) for a clear treatment, using results from Karlin and MacGregor (1972).

14 Kimura (1954) had also discussed the consequences of variation in selection over

time, but viewed this as a mechanism likely to *reduce* polymorphism. See Gillespie (1991, pp. 228–230) for some comments on this part of the history.

15 Gillespie's result is expressed in terms of the probabilistic parameters describing the variation in fitness through time. Haldane and Jayakar's result is expressed in terms of actual sample sequences of fitness relations experienced by the population, not the underlying parameters. This limitation applies to the other Haldane and Jayakar model mentioned in the text also. For temporal variation in selection see also Karlin and Lieberman (1974).

16 A different view of the link between homeostasis and heterozygosity is presented in Gillespie and Langley (1974). They claim that heterozygotes do better in variable environments because of low variance in fitness, but "more homeostatic organisms provide a more buffered and less variable environment" for the enzymes in question. So "analogous genes should be less polymorphic in more homeostatic species" (1974, p. 845).

17 Gillespie's "SAS-CFF" model predicts polymorphism in a very wide range of conditions (Gillespie, 1978, 1991). A special feature of this model is that the fitness of the heterozygote is always closer to the fitness of the *favored* homozygote than it is to the fitness of the other homozygote, in each environmental state. So on the fitness scale, the direction of dominance reverses when the direction of selection reverses. See Hedrick (1986) for a sketch and a comparison to other models.

18 For another model which combines many of the factors discussed so far into a single framework, see Wilson and Yoshimura (1994).

19 I am not going to discuss all the models in Levins' book. The material I will focus on is mainly in Chapter 2, which outlines his basic framework. This is also the material which had the most impact. Many of the models in later chapters are more complicated. See Roughgarden (1979) for a clear step-by-step presentation of Levins' basic model.

20 Levins calls this equal-area principle the "principle of allocation." See Sultan (1992) for a discussion of the role of this principle in subsequent literature.

21 The models in Chapters 7 and 8 used the idea of the "importance" of a state of the world. This is the difference between the payoffs associated with good and bad actions in that state. The importance of environmental state S_3 in Figure 9.2 is the difference between the value of the left hand function at its peak (phenotype = 3) and its value at the point where the other peak is (phenotype = 6), if we assume that producing a phenotype with value 6 is the relevant wrong behavior here.

22 A convex set is a set where a straight line connecting any two points in the set is itself wholly within the set.

23 For a Levins-style model with different optimization principles, see Templeton and Rothman (1974).

24 When bell-shaped functions, or something similar, are used to represent fitness as a function of phenotype, as we are assuming here, the optimum in the case of a concave fitness set is *almost* one of the original optima. In the example we are

discussing it is about 5.99. As the set becomes more concave and the proportions of the environmental states become more asymmetrical, the solution gets closer and closer to one of the original optima.

25 For a good explanation of the technical details, see Roughgarden (1979). The equation for the hyperbolas is: $W_j = (L_G)^{(1/P_j)}(1/(W_i^{(P_i/P_j)}))$.

26 For examples of empirical work which make use of Levins' framework and/or basic concepts, see Selander and Kaufman (1973) and Bryant (1974). A study which makes use both of Levins' ideas and also some of the earlier literature concerning population-level response to environmental variation, discussed in section 9.4, is Marshall and Jain (1968).

27 My discussion in this chapter is much indebted to Seger and Brockmann's presentation. Felsenstein also discusses this problem with Levins' model (1976, p. 257).

28 For the details of models of this type, and the role of geometric mean fitness, see Gillespie (1991), pp. 146–148.

29 $\bar{W}$ is found by averaging the fitnesses of genotypes, weighted by their frequencies. We should note that if population mean fitness in the case in section 9.9 is calculated by averaging the geometric mean fitnesses of genotypes, weighted by the frequencies of the genotypes, then a population comprised solely of I_2 has a higher population mean fitness than the polymorphism of I_1 and I_2 that maximizes L_G.

30 As is more widely recognised, game theory entered evolutionary biology in a similar way. The first paper using it was written by Lewontin (1961) and was group-selectionist in roughly the same sense as Levins' 1968 book. Lewontin's paper includes a discussion of polymorphism as a response to environmental variation. Evolutionary game theory was later developed in an individualistic way by Maynard Smith and others (Maynard Smith and Price, 1973; Maynard Smith, 1982). For a discussion of how some of the issues discussed in this chapter arise in the context of evolutionary game theory, see Maynard Smith (1988) and Bergstrom and Godfrey-Smith (unpublished).

31 Here I bracket the problem raised earlier with Levins' concepts of coarse and fine grain.

References

Allen, C. and M. Bekoff (forthcoming). Function, natural design and animal behavior: philosophical and ethological considerations. To appear in N. Thompson (ed.), *Perspectives in Ethology, Vol. 11.* New York: Plenum.

Amundson, R. (1989). The trials and tribulations of selective explanations. In K. Hahlweg and C. A. Hooker (eds.), *Issues in Evolutionary Epistemology.* State University of New York Press, pp. 413–432.

Antonovics, J., A. D. Bradshaw and R. G. Turner (1971). Heavy metal tolerance in plants. *Advances in Ecological Research* 7: 1–85.

Antonovics, J., N. C. Ellstrand and R. N. Brandon (1988). Genetic variation and environmental variation: expectations and experiments. In L. D. Gottlieb and S. K. Jain (eds.), *Plant Evolutionary Biology.* London: Chapman and Hall, pp. 275–303.

Appiah, A. (1986). Truth conditions: a causal theory. In J. Butterfield (ed.), *Language, Mind and Logic.* Cambridge: Cambridge University Press, pp. 25–45.

Armstrong, D. M. (1968). *A Materialist Theory of the Mind.* London: Routledge and Kegan Paul.

Armstrong, D. M. (1973). *Belief, Truth and Knowledge.* Cambridge: Cambridge University Press.

Ashby, W. R. (1960). *Design for a Brain* (2nd edition). New York: Wiley.

Ashby, W. R. (1963). *An Introduction to Cybernetics.* New York: Wiley.

Axelrod, R. and W. Hamilton (1981). The evolution of cooperation. *Science* 211: 1390–96.

Bain, A. (1859). *Emotions and the Will.* London: Parker.

Beatty, J. (1992). Fitness: theoretical contexts. In Keller and Lloyd (1992), pp. 115–119.

Bechtel, W. and R. C. Richardson (1993). *Discovering Complexity: Decomposition and Localization as Strategies in Scientific Research.* Princeton: Princeton University Press.

291

References

Bergman, A. and M. W. Feldman (in press). On the evolution of learning. *Theoretical Population Biology.*

Bergstrom, C. and P. Godfrey-Smith (unpublished). Evolution of behavioral heterogeneity in individuals and populations.

Berkeley, G. (1710). *Principles of Human Knowledge.* (D. M. Armstrong, ed.) New York: Macmillan, 1965.

Block, N. (1978). Troubles with functionalism. Reprinted in N. Block (ed.), *Readings in the Philosophy of Psychology*, Volume 1. Cambridge MA: Harvard University Press, 1980, pp. 268–305.

Block, N. (1986). Advertisement for a semantics for psychology. Reprinted in Stich and Warfield (1994), pp. 81–141.

Boakes, R. (1984). *From Darwin to Behaviorism.* Cambridge: Cambridge University Press.

Bonner, J. T. (1988). *The Evolution of Complexity.* Princeton: Princeton University Press.

Boorse, C. (1976). Wright on functions. *Philosophical Review* 85: 70–86.

Bowler, P. (1989). *Evolution: The History of an Idea.* (Revised edition) Berkeley: University of California Press.

Bowler, P. (1992). Lamarckism. In Keller and Lloyd (1992), pp. 188–193.

Boyd, R. (1983). On the current status of scientific realism. *Erkenntnis* 19: 45–90. Reprinted (revised) in Boyd, Gaspar and Trout (1991).

Boyd, R., P. Gasper and J. D. Trout (eds.) (1991). *The Philosophy of Science.* Cambridge MA: MIT Press, 1991.

Bradie, M. (1994). Epistemology from an evolutionary point of view. In E. Sober (ed.), *Conceptual Issues in Evolutionary Biology.* 2nd edition. Cambridge MA: MIT Press, pp. 453–475.

Bradshaw, A. D. (1965). Evolutionary significance of phenotypic plasticity in plants. *Advances in Genetics* 13: 115–55.

Braitenberg, V. (1984). *Vehicles: Experiments in Synthetic Psychology.* Cambridge MA: MIT Press.

Brandon, R. N. (1990). *Adaptation and Environment.* Princeton: Princeton University Press.

Brandon, R. N. and J. Antonovics. (forthcoming). The coevolution of organism and environment. To appear in *Proceedings of the Pittsburgh-Konstanz Philosophy of Biology Colloquium.*

Broad, C. D. (1925). *The Mind and Its Place in Nature.* London: Routledge and Kegan Paul.

Brönmark, C. and J. G. Miner (1992). Predator-induced phenotypical change in body morphology in crucian carp. *Science* 258: 1348–1350.

Bryant, E. (1974). On the adaptive significance of enzyme polymorphisms in relation to environmental variability. *American Naturalist* 108: 1–19.

Burge, T. (1979). Individualism and the mental. In P. A. French, T. E. Uehling an H. K.

References

Wettstein (eds.), *Midwest Studies in Philosophy IV: Studies in Metaphysics*. Minneapolis: University of Minnesota Press, pp. 73–121.

Burian, R. (1983). Adaptation. In M. Grene (ed.), *Dimensions of Darwinism*. Cambridge: Cambridge University Press, pp. 287–314.

Burian, R. (1992). Adaptation: historical perspectives. In Keller and Lloyd (1992), pp. 7–12.

Burke, T. (1994). *Dewey's New Logic*. Chicago: Chicago University Press.

Burkhardt, R. (1977). *The Spirit of System: Lamarck and Evolutionary Biology*. Cambridge MA: Harvard University Press.

Byerly, H. C. and R. E. Michod (1991). Fitness and evolutionary explanation. *Biology and Philosophy* 6: 1–22.

Byrne, R. W. and A. Whiten (eds.) (1988). *Machiavellian Intelligence: Social Expertise and the Evolution of Intellect in Monkeys and Humans*. Oxford: Clarendon Press.

Campbell, D. T. (1974). Evolutionary epistemology. In *The Philosophy of Karl Popper*. (P. A. Schillp, ed.) La Salle: Open Court, pp. 413–63.

Cannon, W. B. (1932). *The Wisdom of the Body*. New York: Norton.

Cavalli-Sforza, L. L. and W. F. Bodmer (1971). *The Genetics of Human Populations*. San Francisco: Freeman.

Changeaux, J. P. (1985). *Neuronal Man: The Biology of Mind*. (L. Garey, trans.) New York: Pantheon.

Chomsky, N. (1956). Review of B. F. Skinner's *Verbal Behavior*. Reprinted (abridged) in N. Block (ed.), *Readings in the Philosophy of Psychology*, Volume 1. Cambridge MA: Harvard University Press, 1980, pp. 48–63.

Chomsky, N. (1980). *Rules and Representations*. New York: Columbia University Press.

Chomsky, N. (1988). *Language and Problems of Knowledge: The Managua Lectures*. Cambridge MA: MIT Press.

Christiansen, F. B. (1975). Hard and soft selection in a subdivided population. *American Naturalist* 109: 11–16.

Christiansen, F. B. and M. Feldman (1975). Subdivided populations: a review of the one- and two-locus theory. *Theoretical Population Biology* 7: 13–38.

Clark, C. W. and C. D. Harvell (1991). Inducible defences and the allocation of resources: a minimal model. *American Naturalist* 139: 521–39.

Cohen, D. (1967). Optimizing reproduction in a randomly varying environment when a correlation exists between the conditions at the time a choice has to be made and the subsequent outcome. *Journal of Theoretical Biology* 16: 1–14.

Coleman, W. (1971). *Biology in the Nineteenth Century*. Cambridge: Cambridge University Press.

Conrad, M. (1983). *Adaptability: The Significance of Variability from Molecule to Ecosystem*. New York: Plenum.

Coombs, C. H., R. M. Dawes and A. Tversky (1970). *Mathematical Psychology: An Elementary Introduction*. Englewood Cliffs: Prentice-Hall.

References

Cooper, W. S. and R. H. Kaplan (1982). Adaptive "coin-flipping": a decision-theoretic examination of natural selection for random individual variation. *Journal of Theoretical Biology* 94: 135–51.

Cosmides, L. and J. Tooby (1987). From evolution to behavior: evolutionary psychology as the missing link. In Dupre (1987), pp. 277–306.

Cummins, R. (1975). Functional analysis. *Journal of Philosophy* 72: 741–764.

Cuvier, G. (1827). *Essay on the Theory of the Earth.* (3rd edition, translated from French. Additional material by R. Jameson). Edinburgh: William Blackwood.

Curtis, H. and N. S. Barnes (1989). *Biology.* (5th edition) New York: Worth.

Dahlbom, B. (1985). Dennett on cognitive ethology: a broader view. *Behavioral and Brain Sciences* 8: 760–61.

Darwin, C. (1859). *On The Origin of Species by Means of Natural Selection: Or the Preservation of Favoured Races in the Struggle for Life.* Reprinted with introduction by E. Mayr. Cambridge MA: Harvard University Press, 1964.

Darwin, C. (1868). *The Variation of Animals and Plants Under Domestication.* 2 volumes. London: John Murray.

Darwin, C. (1969). *The Autobiography of Charles Darwin, 1809–1882.* (N. Barlow, ed.) New York: Norton, 1969.

Davidson, D. (1975). Thought and talk. Reprinted in Davidson (1984), pp. 155–170.

Davidson, D. (1984). *Essays on Truth and Interpretation.* Oxford: Oxford University Press.

Dawkins, R. (1976). *The Selfish Gene.* Oxford: Oxford University Press.

Dawkins, R. (1982). *The Extended Phenotype.* Oxford: Oxford University Press.

Dawkins, R. (1986). *The Blind Watchmaker.* New York: Norton.

Dempster, E. (1955). Maintenance of genetic diversity. *Cold Spring Harbour Symposium on Quantitative Biology* 20: 25–31.

Dempster, E. (1956). Comments on Professor Lewontin's article. *American Naturalist* 90: 385–386.

Dennett, D. C. (1975). Why the Law of Effect will not go away. Reprinted in Dennett (1978), pp. 71–89.

Dennett, D. C. (1978). *Brainstorms: Philosophical Essays on Mind and Psychology.* Cambridge MA: MIT Press.

Dennett, D. C. (1980). Passing the buck to biology. *Behavioral and Brain Sciences* 3: 19.

Dennett, D. C. (1981). Three kinds of intentional psychology. Reprinted in Dennett (1987), pp. 43–68.

Dennett, D. C. (1983). Intentional systems in cognitive ethology: the "Panglossian" paradigm defended. *Behavioral and Brain Sciences* 6: 343–355.

Dennett, D. C. (1987). *The Intentional Stance.* Cambridge MA: MIT Press.

Dennett, D. C. (1995). *Darwin's Dangerous Idea: Evolution and the Meanings of Life.* New York: Simon and Schuster.

Devitt, M. (1991a). *Realism and Truth.* Second edition. Oxford: Blackwell.

References

Devitt, M. (1991b). Minimalist truth: a critical notice of Paul Horwich's *Truth*. *Mind and Language* 6: 273–283.

Devitt, M. and K. Sterelny (1989). *Language and Reality*. Cambridge MA: MIT Press.

Dewey, J. (1898). Evolution and ethics. *The Monist* 8: 321–41.

Dewey, J. (1904). The philosphical work of Herbert Spencer. *Philosophical Review* 13: 159–75. Reprinted in *John Dewey: The Middle Works, 1899–1924. Volume 3: 1903–1906* (J. A. Boydston, ed.) Carbondale: Southern Illinois University Press, 1977, pp. 193–209

Dewey, J. (1910a). A short catechism concerning truth. Reprinted in Dewey (1985), pp. 3–11.

Dewey, J. (1910b). The influence of Darwin on philosophy. Reprinted in Dewey (1981), pp. 31–41.

Dewey, J. (1911a). Adaptation. In *A Cyclopedia of Education*. (Paul Monroe, ed.) New York: Macmillan, 1911. Reprinted in Dewey (1985), pp. 364–366.

Dewey (1911b). Organism and environment. In *A Cyclopedia of Education*. (Paul Monroe, ed.) New York: Macmillan, 1911. Reprinted in Dewey (1985), pp. 437–440.

Dewey, J. (1917). The need for a recovery of philosophy. Reprinted in Dewey (1981), pp. 58–97.

Dewey, J. (1922). *Human Nature and Conduct*. New York: Henry Holt.

Dewey, J. (1929a). *Experience and Nature* (revised edition). New York: Dover, 1958. (First edition 1925.)

Dewey, J. (1929b). *The Quest for Certainty*. Reprinted in *John Dewey: The Later Works, 1925–1953. Volume 4: 1929*. (J. A. Boydston, ed.) Carbondale: Southern Illinois University Press, 1988.

Dewey, J. (1931). The practical character of reality. Reprinted in Dewey (1985), pp. 207–222.

Dewey, J. (1934). *Art as Experience*. Reprinted in *John Dewey: The Later Works, 1925–1953. Volume 10: 1934*. (J. A. Boydston, ed.) Carbondale: Southern Illinois University Press, 1989.

Dewey, J. (1938). *Logic: The Theory of Inquiry*. Reprinted in *John Dewey: The Later Works, 1925–1953. Volume 12: 1938*. (J. A. Boydston, ed.) Carbondale: Southern Illinois University Press, 1991.

Dewey, J. (1941). Propositions, warranted assertibility, and truth. *Journal of Philosophy* 38: 169–86.

Dewey, J. (1948). *Reconstruction in Philosophy*. Enlarged edition. Boston: Beacon. (First edition 1920.)

Dewey, J. (1981). *The Philosophy of John Dewey*, (J. J. McDermott, ed.) Chicago: Chicago University Press.

Dewey, J. (1985). *John Dewey: The Middle Works, 1899–1924. Volume 6. How We Think, and Selected Essays, 1910–1911*. (J. A. Boydston, ed.) Carbondale: Southern Illinois University Press.

References

Dewey, J. and Bentley, A. F. (1949). *Knowing and the Known*. Reprinted in *John Dewey: The Later Works, 1925–1953, Volume 16: 1949–1952*. (J. A. Boydston, ed.) Carbondale: Southern Illinois University Press, 1989.

Diderot, D. (1774). Excerpt from *Refutation of Helvetius* (C. Brinton, trans.) In *The Portable Age of Reason*. (C. Brinton, ed.) New York: Viking, pp. 263–265.

Dobzhansky, T. (1951). *Genetics and the Origin of Species*, 3rd edition. New York: Columbia University Press.

Dobzhansky, T. (1955). A review of some fundamental concepts and problems of population genetics. *Cold Springs Harbor Symposium in Quantitative Biology* 20: 1–15.

Dobzhansky, T. and H. Levene (1955). Developmental homeostasis in natural populations of *Drosophila Pseudoobscura*. *Genetics* 40: 797–808.

Dobzhansky, T. and B. Spassky (1944). Manifestation of genetic variants in *Drosophila Pseudoobscura* in different environments. *Genetics* 29: 270–290.

Dobzhansky, T. and B. Wallace (1953). The genetics of homeostasis in *Drosophila*. *Proceedings of the National Academy of Sciences, USA* 39: 162–171.

Dobzhansky, T. et al. (1981). *Dobzhansky's Genetics of Natural Populations, I-XLIII*. (R. C. Lewontin, J. A. Moore, W. B. Provine and B. Wallace, eds.) New York: Columbia University Press.

Dretske, F. (1981). *Knowledge and the Flow of Information*. Cambridge MA: MIT Press.

Dretske, F. (1983). Precis of Dretske (1981), with commentaries and replies. *Behavioral and Brain Sciences* 6: 55–63.

Dretske, F. (1988). *Explaining Behavior*. Cambridge MA: MIT Press.

Dupre, J. (1987). *The Latest on the Best. Essays on Evolution and Optimality*. Cambridge MA: MIT Press.

Eshel, J. E. and M. W. Feldman (1982). On evolutionary genetic stability of the sex ratio. *Theoretical Population Biology* 21: 431–39.

Felsenstein, J. (1976). The theoretical population genetics of variable selection and migration. *Annual Review of Genetics* 19: 253–80.

Field, H. (1972). Tarski's theory of truth. *Journal of Philosophy* 69: 347–75.

Field, H. (1986). The deflationary conception of truth. In G. MacDonald and C. Wright (eds.) *Fact, Science and Morality*. Oxford: Blackwell, pp. 55–117.

Field, H. (1990). "Narrow" aspects of intentionality and the information-theoretic approach to content. In E. Villanueva (ed.) *Information, Semantics and Epistemology*. Oxford: Blackwell.

Fodor, J. A. (1975). *The Language of Thought*. New York: Crowell.

Fodor, J. A. (1980). Methodological solipsism considered as a research strategy in cognitive psychology. Reprinted in Fodor (1981), pp. 225–253.

Fodor, J. A. (1981). *Representations*. Cambridge MA: MIT Press.

Fodor, J. A. (1984). Semantics, Wisconsin style. *Synthese* 59: 1–20.

Fodor, J. A. (1987). *Psychosemantics*. Cambridge MA: MIT Press.

Fodor, J. A. (1990). *A Theory of Content and Other Essays*. Cambridge MA: MIT Press.

References

Ford, E. B. (1975). *Ecological Genetics*, 4th edition. New York: John Wiley.

Francis, R. (1990). Causes, proximate and ultimate. *Biology and Philosophy* 5: 401–15.

Francis, R. (1992). Sexual lability in teleosts: developmental factors. *Quarterly Review of Biology* 67: 1–18.

Francis, R. and G. W. Barlow (1993). Social control of primary sex determination in the Midas cichlid. *Proceedings of the National Academy of Sciences, USA* 90: 10673-10675.

Francis, R. (in preparation). Of shrimps and chimps.

Futuyma, D. J. and G. Moreno (1988). The evolution of ecological specialization. *Annual Review of Ecology and Systematics* 19: 207–33.

Galison, P. (1990). Aufbau/Bauhaus: logical positivism and architectural modernism. *Critical Inquiry* 16: 709–752.

Gibson, J. J. (1966). *The Senses Considered as Perceptual Systems*. Boston: Houghton Mifflin.

Gibson, J. J. (1979). *The Ecological Approach to Visual Perception*. Boston: Houghton Mifflin.

Gillespie, J. H. (1973). Polymorphism in random environments. *Theoretical Population Biology* 4: 193–195.

Gillespie, J. H. (1978). A general model to account for enzyme variation in natural populations V. The SAS-CFF model. *Theoretical Population Biology* 14: 1–45.

Gillespie, J. H. (1991). *The Causes of Molecular Evolution*. New York: Oxford University Press.

Gillespie, J. H. and C. H. Langley (1974). A general model to account for enzyme variation in natural populations. *Genetics* 76: 837–884.

Godfrey-Smith, P. (1991). Signal, decision, action. *Journal of Philosophy* 88: 709–722.

Godfrey-Smith, P. (1992). Indication and adaptation. *Synthese* 92: 283–312.

Godfrey-Smith, P. (1993). Functions: consensus without unity. *Pacific Philosophical Quarterly* 74: 196–208.

Godfrey-Smith, P. (1994a). A modern history theory of functions. *Nous* 28: 344–362.

Godfrey-Smith, P. (1994b). A continuum of semantic optimism. In Stich and Warfield (1994), pp. 259–277.

Godfrey-Smith, P. (1994c). Spencer and Dewey on life and mind. In R. Brooks and P. Maes (eds.) *Artificial Life 4*. Cambridge MA: MIT Press, pp. 80–89.

Godfrey-Smith, P. (unpublished). The varieties of reliability.

Goldman, A. (1976). Discrimination and perceptual knowledge. *Journal of Philosophy* 73: 771–791.

Goldman, A. (1986). *Epistemology and Cognition*. Cambridge MA: Harvard University Press.

Good, I. J. (1967). On the principle of total evidence. *British Journal for the Philosophy of Science* 17: 310–21.

Goodman, N. (1955). *Fact, Fiction and Forecast*. Indianapolis: Bobbs-Merrill.

Goodwin, B. (1996). *How the Leopard Changed its Spots: The Evolution of Complexity*. New York: Touchstone.

References

Gordon, D. (1991). Variation and change in behavioral ecology. *Ecology* 72: 1196–1203.

Gordon, D. (1992). Phenotypic plasticity. In Keller and Lloyd (1992), pp. 255–262.

Goudge, T. A. (1973). Pragmatism's contribution to an evolutionary view of mind. *Monist* 57: 133–50.

Gould, S. J. (1977). Eternal metaphors of paleontology. In A. Hallam (ed.) *Patterns of Evolution as Illustrated by the Fossil Record*. Amsterdam: Elsevier, 1977, pp. 1–26.

Gould, S. J. and R. C. Lewontin (1978). The spandrels of San Marco and the panglossian paradigm: a critique of the adaptationist program. *Proceedings of the Royal Society, London* 205: 581–598.

Gowen, J. W. (ed.) (1964). *Heterosis*. New York: Hafner.

Gray, R. (1992). The death of the gene. In Griffiths (1992b), pp. 165–209.

Green, D. M. and J. A. Swets (1966). *Signal Detection and Psychophysics*. New York: John Wiley and Sons.

Griffiths, P. (1992a). Adaptive explanation and the concept of a vestige. In Griffiths (1992b), pp. 111–32.

Griffiths, P. (ed.) (1992b). *Trees of Life: Essays in Philosophy of Biology*. Dordrecht: Kluwer.

Hacking, I. (1983). *Representing and Intervening*. Cambridge: Cambridge University Press.

Haldane, J. B. S. and S. D. Jayakar (1963). Polymorphism due to selection of varying direction. *Journal of Genetics* 58: 237–242.

Hamilton, W. D. (1964). The genetical evolution of social behavior, I and II. *Journal of Theoretical Biology* 7: 1–52.

Harvell. D. (1986). The ecology and evolution of inducible defences in a marine bryozoan: cues, costs, and consequences. *American Naturalist* 128: 810–23.

Hedrick, P. W., M. E. Ginevan, and E. P. Ewing (1976). Genetic polymorphism in heterogeneous environments. *Annual Review of Ecology and Systematics* 7: 1–32.

Hedrick, P. W. (1986). Genetic polymorphism in heterogeneous environments: a decade later. *Annual Review of Ecology and Systematics* 17: 535–66.

Heilbroner, R. L. (1992). *The Worldly Philosophers*. 6th edition. New York: Simon and Schuster.

Helvétius, C. A. (1773). *A Treatise on Man*. (W. Hooper, trans.) London, 1877.

Himmelfarb, G. (1959). *Darwin and the Darwinian Revolution*. New York: Doubleday.

Horwich, P. (1990). *Truth*. Oxford: Blackwell.

Huey, R. B. and P. E. Hertz (1984). Is a jack-of-all-temperatures a master of none? *Evolution* 38: 441–444.

Hull, D. L. (1988). *Science as a Process*. Chicago: University of Chicago Press.

Hume, D. (1739). *A Treatise of Human Nature*. (L. A. Selby-Bigge, ed.) 2nd edition. Oxford: Clarendon Press, 1978.

Hume, D. (1777). *An Enquiry Concerning the Human Understanding and Concerning the*

References

Principles of Morals. (L. A. Selby-Bigge, ed.) 2nd edition. Oxford: Clarendon Press, 1962.

Humphrey, N. (1976). The social function of intellect. Reprinted in Byrne and Whiten (1988), pp. 13–26.

James, W. (1878). Remarks on Spencer's definition of mind as Correspondence. *Journal of Speculative Philosophy* 12: 1–18. Reprinted in *William James: The Essential Writings* (B. Wilshire, ed.). New York: Harper and Row, 1971, pp. 9–24.

James, W. (1880). Great men and their environments. Reprinted in James (1897b), pp. 216–254.

James, W. (1890). *Principles of Psychology.* 2 volumes. New York: Henry Holt.

James, W. (1896). The will to believe. Reprinted in James (1897b), pp. 1–31.

James, W. (1897a). The sentiment of rationality. In James (1897b), pp. 63–110.

James, W. (1897b). *The Will to Believe, and Other Essays in Popular Philosophy.* New York: Longmans.

James, W. (1898). Philosophical conceptions and practical results. Reprinted in *The Writings of William James* (J. J. McDermott, ed.) Chicago: Chicago University Press, 1977, pp. 345–362.

James, W. (1907). *Pragmatism: A New Name for Some Old Ways of Thinking.* NewYork: Longmans. Reprinted in James (1975).

James, W. (1909). Humanism and truth. Reprinted in James (1975), pp. 203–26.

James, W. (1975). *Pragmatism and The Meaning of Truth.* Cambridge MA: Harvard University Press.

Jeffrey, R. (1983). *The Logic of Decision.* 2nd edition. Chicago: Chicago University Press.

Kant, I. (1781). *Critique of Pure Reason.* (trans. N. Kemp Smith) London: Macmillan, 1929.

Karlin, S. and U. Lieberman (1974). Random temporal variation in selection intensities: Case of large population size. *Theoretical Population Biology* 6: 355–82.

Karlin, S. and J. L. MacGregor (1972). Polymorphisms for genetic and ecological systems with weak coupling. *Theoretical Population Biology* 3: 210–38.

Kauffman, S. (1993). *The Origins of Order.* Oxford: Oxford University Press.

Keller, E. F. and E. A. Lloyd (eds.) (1992). *Keywords in Evolutionary Biology.* Cambridge: Harvard University Press.

Kendall, M. G. and A. Stuart (1961). *The Advanced Theory of Statistics, Volume II: Inference and Relationship.* New York: Hafner.

Kennedy, J. G. (1978). *Herbert Spencer.* Boston: Twayne.

Kettlewell, H. B. D. (1973). *The Evolution of Melanism.* Oxford: Oxford University Press.

Kimura, M. (1954). Processes leading to quasi-fixation of alleles in natural populations due to random fluctuations in selection intensities. *Genetics* 39: 280–95.

Kimura, M. (1983). *The Neutral Theory of Molecular Evolution.* Cambridge: Cambridge University Press.

References

Kimura, M. (1991). Recent development of the neutral theory viewed from the Wrightian tradition of theoretical population genetics. *Proceedings of the National Academy of Science, USA* 88: 5969–5973.

Kitcher, P. S. (1985). *Vaulting Ambition; Sociobiology and the Quest for Human Nature.* Cambridge MA: MIT Press.

Kitcher, P. S. (1992). The naturalist's return. *Philosopical Review* 101: 53–113.

Kitcher, P. S. (1993). The evolution of human altruism. *Journal of Philosophy* 90: 497–516.

Kitcher, P. S. (1993). Function and design. In *Midwest Studies in Philosophy, Vol. 18* (P. A. French, T. E. Uehling, Jr. and H. K. Wettstein, eds.). Notre Dame: University of Notre Dame Press, pp. 379–397.

Kojima, K. and T. M. Kelleher (1961). Changes of mean fitness in randomly mating populations when epistasis and linkage are present. *American Naturalist* 46: 527–540.

Kornblith, H. (ed.) (1985). *Naturalizing Epistemology.* Cambridge MA: MIT Press.

Krebs, J. and N. Davies (1987). *An Introduction to Behavioural Ecology.* 2nd edition. Oxford: Blackwell.

Kuhn, T. (1962). *The Structure of Scientific Revolutions.* Chicago: Chicago University Press.

Lamarck, J. B. (1809). *Zoological Philosophy* (Hugh Elliot, trans.). Chicago: University of Chicago Press, 1984.

Langton, C. (ed.) (1989). *Artificial Life. SFI Studies in the Sciences of Complexity.* Reading, MA: Addison-Wesley.

Langton, C. G., C. Taylor, J. D. Farmer, S. Rasmussen, (eds.) (1992). *Artificial Life II. SFI Studies in the Sciences of Complexity, Vol. X.* Reading, MA: Addison-Wesley.

Langton, C. G. (1992). Life at the edge of chaos. In Langton et al. (1992), pp. 41–89.

Latour, B. (1987). *Science in Action.* Cambridge: Harvard University Press.

Latour, B. (1992). One more turn after the social turn. In *The Social Dimensions of Science* (E. McMullin, ed.). Notre Dame: Notre Dame University Press, pp. 272–94.

Latour, B. and S. Woolgar (1986). *Laboratory Life* (revised edition). Princeton: Princeton University Press. (First edition 1979).

Lehrer, K. (1974). *Knowledge.* Oxford: Clarendon Press.

Leibniz, G. W. (1765). *New Essays on Human Understanding.* (P. Remnant and J. Bennett, trans. and ed.) Cambridge: Cambridge University Press, 1981.

Lenoir, T. (1982). *The Strategy of Life: Teleology and Mechanism in Nineteenth-century German Biology.* Chicago: Chicago University Press.

Lerner, I. M. (1954). *Genetic Homeostasis.* Edinburgh: Oliver and Boyd.

Levene, H. (1953). Genetic equilibrium when more than one ecological niche is available. *American Naturalist* 87: 331–33.

Levi, I. (1967). *Gambling with Truth.* New York: Alfred A. Knopf.

Levi, I. (1981). *The Enterprise of Knowledge.* Cambridge MA: MIT Press.

References

Levins, R. (1962). Theory of fitness in a heterogeneous environment. I. The fitness set and adaptive function. *American Naturalist* 96: 361–373.

Levins, R. (1963). Theory of fitness in a heterogeneous environment. II. Developmental flexibility and niche selection. *American Naturalist* 97: 75–90.

Levins, R. (1965). Theory of fitness in a heterogeneous environment V. Optimal genetic systems. *Genetics* 52: 891–904.

Levins, R. (1968a). *Evolution in Changing Environments*. Princeton: Princeton University Press.

Levins, R. (1968b). Evolutionary consequences of flexibility. In R. Lewontin (ed.) *Population Biology and Evolution*. Syracuse: Syracuse University Press, pp. 67–70.

Levins, R. (1979). Coexistence in a variable environment. *American Naturalist* 114: 765–783.

Levins, R. and R. C. Lewontin (1985). *The Dialectical Biologist*. Cambridge MA: Harvard University Press.

Levins, R. and R. MacArthur (1966). The maintenance of polymorphism in a spatially heterogeneous environment: variations on a theme by Howard Levene. *American Naturalist* 100: 585–89.

Lewin, R. (1992). *Complexity: Life at the Edge of Chaos*. New York: Macmillan.

Lewis, D. K. (1972). Psychophysical and theoretical identifications. Reprinted in Block (1980), pp. 207–215.

Lewis, D. K. (1980). Mad pain and Martian pain. In Block (1980), pp. 216–222.

Lewis, D. K. (1983). New work for a theory of universals. *Australasian Journal of Philosophy* 61: 434–77.

Lewontin, R. C. (1956). Studies on homeostasis and heterozygosity I. General considerations. Abdominal bristle number in second chromosome homozygotes of *Drosophila Melanogaster*. *American Naturalist* 90: 237–255.

Lewontin, R. C. (1957). The adaptations of populations to varying environments. *Cold Spring Harbour Symposium on Quantitative Biology* 22: 395–408.

Lewontin, R. C. (1958). Studies on heterozygosity and homeostasis II: loss of heterosis in a constant environment. *Evolution* 12: 494–503.

Lewontin, R. C. (1961). Evolution and the theory of games. *Journal of Theoretical Biology* 1: 382–402.

Lewontin, R. C. (1970). The units of selection. *Annual Review of Ecology and Systematics* 1: 1–18.

Lewontin, R. C. (1974a). *The Genetic Basis of Evolutionary Change*. New York: Columbia.

Lewontin, R. C. (1974b). The analysis of variance and the analysis of causes. Reprinted in Levins and Lewontin (1985), pp. 109–22.

Lewontin, R. C. (1977a). Evolution as theory and ideology. First published in Italian in Giulio Enaudi (ed.), *Enciclopedia Enaudi*, volume 3. Published (abridged) in English in Levins and Lewontin (1985), pp. 9–64.

Lewontin, R. C. (1977b). Adaptation. First published in Italian in Giulio Enaudi (ed.)

References

Enciclopedia Enaudi, volume 1. Published in English in Levins and Lewontin (1985), pp. 65–84.

Lewontin, R. C. (1982). Organism and environment. In Plotkin (1982), pp. 151–170.

Lewontin, R. C. (1983). The organism as the subject and object of evolution. Reprinted in Levins and Lewontin (1985), pp. 85–106.

Lewontin, R. C. (1991). *Biology as Ideology: The Doctrine of DNA*. New York: Harper.

Li, C. C. (1955). The stability of an equilibrium and the average fitness of a population. *American Naturalist* 89: 281–95.

Lively, C. (1986). Canalization versus developmental conversion in a spatially variable environment. *American Naturalist* 128: 561–72.

Lloyd, E. A. (1988). *The Structure and Confirmation of Evolutionary Theory*. New York: Greenwood Press.

Locke, J. (1690). *An Essay Concerning Human Understanding.* (A. C. Fraser, ed.) 2 volumes. New York: Dover, 1959.

Lycan, W. G. (1981). Form, function and feel. *Journal of Philosophy* 78: 24–50.

MacArthur, R. H. (1972). *Geographical Ecology*. New York: Harper and Row.

MacKenzie, D. (1981). *Statistics in Britain, 1865–1930*. Edinburgh: Edinburgh University Press.

Marshall, D. R. and S. K. Jain (1968). Phenotypic plasticity of *Avena Fatua* and *A. Barbata. American Naturalist* 102: 457–67.

Maturana, H. (1970). Biology of cognition. Reprinted in Maturana and Varela (1980), pp. 1–58.

Maturana, H. and F. Varela (1973). Autopoiesis: the organization of the living. Reprinted in Maturana and Varela (1980), pp. 59–140.

Maturana, H. and F. Varela (1980). *Autopoiesis and Cognition*. Boston Studies in the Philosophy of Science. Dordrecht: Reidel.

Maynard Smith, J. (1970). Genetic polymorphism in a varied environment. *American Naturalist*: 104: 487–490.

Maynard Smith, J. (1982). *Evolution and the Theory of Games*. Cambridge: Cambridge University Press.

Maynard Smith, J. (1988). Can a mixed strategy be stable in a finite population? *Journal of Theoretical Biology* 130: 247–51.

Maynard Smith, J. (1993). *The Theory of Evolution*, 3rd edition. Cambridge University Press.

Maynard Smith, J, R. Burian, S. Kauffman, P. Alberch, J. Campbell, B. Goodwin, R. Lande, D. Raup and L. Wolpert (1985). Developmental constraints and evolution. *Quarterly Review of Biology* 60: 265–287.

Maynard Smith, J. and G. R. Price (1973). The logic of animal conflict. *Nature* 246: 15–18.

Mayr, E. (1961). Cause and effect in biology. *Science* 134: 1501–1506.

Mayr, E. (1974). Behavior programs and evolutionary strategies. *American Scientist* 62: 650–659.

References

Mayr, E. (1982). *The Growth of Biological Thought.* Cambridge: Harvard University Press.

McMullin, E. (1988). *Construction and Constraint: The Shaping of Scientific Rationality.* Notre Dame: University of Notre Dame Press.

McShea, D. (1991). Complexity and evolution: what everybody knows. *Biology and Philosophy* 6: 303–24.

Meyer, J-A, and S. W. Wilson (eds.) (1991). *From Animals to Animats: Proceedings of the First International Conference on the Simulation of Adaptive Behavior.* Cambridge MA: MIT Press.

Millikan, R.G. (1984). *Language, Thought, and Other Biological Categories.* Cambridge MA: MIT Press.

Millikan, R. G. (1986). Thoughts without laws; cognitive science with content. *Philosophical Review* 95: 47–80.

Millikan, R. G. (1989a). In defence of proper functions. *Philosophy of Science* 56: 288–302.

Millikan, R. G. (1989b). Biosemantics. *Journal of Philosophy* 86: 281–297.

Millikan, R. G. (1989c). An ambiguity in the notion "function." *Biology and Philosophy* 4: 172–176.

Mills, S. and J. Beatty (1979). The propensity interpretation of fitness. *Philosophy of Science* 46: 263–288.

Moore, G. E. (1907). William James' "Pragmatism." Reprinted in *Philosophical Studies.* London: Routledge and Kegan Paul, 1922, pp. 97–146.

Moran, N. (1992). The evolutionary maintenance of alternative phenotypes. *American Naturalist* 139: 971–89.

Neander, K. (1991). The teleological notion of "function." *Australasian Journal of Philosophy* 69: 454–468.

Neyman, J. (1950). *First Course in Probability and Statistics.* New York: Henry Holt.

Nozick, R. (1981). *Philosophical Explanations.* Cambridge MA: Harvard University Press.

Orzack, S. (1985). Population dynamics in variable environments V. The genetics of homeostasis revisited. *American Naturalist* 125: 550–572.

Orzack, S. H. (1990). The comparative biology of sex ratio evolution within a natural population of a parasitic wasp *Nasonia Vitripennis. Genetics* 124: 385–396.

Orzack, S. H. and E. Sober (1994). Optimality models and the test of adaptationism. *American Naturalist* 143: 361–380.

Oyama, S. (1985). *The Ontogeny of Information.* Cambridge: Cambridge University Press.

Oyama, S. (1992). Ontogeny and phylogeny: a case of metarecapitulation? In Griffiths (1992b), pp. 211–239.

Paley, W. (1802). *Natural Theology: Or Evidences of the Existence and Attributes of the Deity Collected from the Appearances of Nature.* London: reprinted Farnborough, Gregg, 1970.

References

Papineau, D. (1987). *Reality and Representation*. Oxford: Blackwell.

Paul, D. (1992a). Fitness: historical perspectives. In Keller and Lloyd (1992), pp. 112–114.

Paul, D. (1992b). Heterosis. In Keller and Lloyd (1992), pp. 166–69.

Peel, J. D. Y. (1971). *Herbert Spencer: The Evolution of a Sociologist*. New York: Basic Books.

Peirce, C. S. (1877). The fixation of belief. First published in *Popular Science Monthly*. Reprinted in Peirce (1955), pp. 5–22.

Peirce, C. S. (1878). How to make our ideas clear. First published in *Popular Science Monthly*. Reprinted in Peirce (1955), pp. 23–41.

Peirce, C. S. (1955). *Philosophical Writings* (J. Buchler, ed.). New York: Dover.

Perry, R. B. (1935). *The Thought and Character of William James*. Boston: Little, Brown and Co.

Phillips, D. (1971). John Dewey and the organismic archetype. In *Melbourne Studies in Education 1971* (R. J. W. Selleck, ed.) Melbourne: Melbourne University Press, pp. 171–232.

Piaget, J. (1971). *Biology and Knowledge*. (B. Walsh, trans.) Chicago: Chicago University Press.

Piatelli-Palmerini, M. (1989). Evolution, selection and cognition: from 'learning' to parameter setting in biology and the study of language. *Cognition* 31: 1–44.

Pietroski, P. (1992). Intentionality and teleological error. *Pacific Philosophical Quarterly* 73: 267–82.

Pittendrigh, C. S. (1958). Adaptation, natural selection, and behavior. In *Behavior and Evolution*. (A. Roe and G. G. Simpson, eds.) New Haven: Yale University Press.

Plotkin, H. C. (ed.) (1982). *Learning, Development and Culture: Essays in Evolutionary Epistemology*. New York: Wiley.

Plotkin, H. C. and F. J. Odling-Smee (1979). Learning, change and evolution: an inquiry into the teleonomy of learning. *Advances in the Study of Behavior* 10: 1–41.

Popper, K. R. (1959). *The Logic of Scientific Discovery*. London: Hutchinson.

Popper, K. (1963). Science: conjectures and refutations. Reprinted in *Conjectures and Refutations: The Growth of Scientific Knowledge*. Fifth edition. London: Routledge and Kegan Paul, 1974, pp. 33–65.

Popper, K. R. (1972). Of clouds and clocks. Reprinted in Plotkin (1982), pp. 109–119.

Popper, K. R. (1974). Autobiography. In *The Philosophy of Karl Popper* (P. A. Schillp, ed.). La Salle: Open Court, pp. 3–181.

Prout, T. (1968). Sufficient conditions for multiple-niche polymorphism. *American Naturalist* 102: 493–96.

Provine, W. (1971). *The Rise of Theoretical Population Genetics*. Chicago: Chicago University Press.

Putnam, H. (1960). Minds and Machines. In S. Hook (ed.) *Dimensions of Mind*. New York: New York University Press, pp. 148–179.

Putnam, H. (1975). The meaning of 'meaning.' Reprinted in *Mind, Language and*

References

Reality: Philosophical Papers, Vol. 2. Cambridge: Cambridge University Press, 1975, pp. 215–271.

Putnam, H. (1978). *Meaning and the Moral Sciences.* London: Routledge and Kegan Paul.

Quine, W. V. (1969). Epistemology naturalized. Reprinted in Kornblith (1985), pp. 31–47.

Ramsey, F. P. (1927). Facts and propositions. In *The Foundations of Mathematics, and Other Logical Essays,* (R. B. Braithwaite, ed.) London: Routledge and Kegan Paul, 1931, pp. 138–55.

Ray, J. (1692). *The Wisdom of God Manifested in the Works of the Creation.* 2nd edition. London: Samuel Smith. First edition 1691.

Rescher, N. (1977). *Methodological Pragmatism.* New York: New York University Press.

Richards, R. (1987). *Darwin and the Emergence of Evolutionary Theories of Mind and Behavior.* Chicago: University of Chicago Press.

Robertson, F. W. and E. C. R. Reeve (1952). Heterozygosity, environmental variation and heterosis. *Nature* 170: 286.

Rorty, R. (1977). Dewey's metaphysics. Reprinted in Rorty (1982), pp. 72–89.

Rorty, R. (1979). *Philosophy and the Mirror of Nature.* Princeton: Princeton University Press.

Rorty, R. (1982). *Consequences of Pragmatism.* Minneapolis: University of Minnesota Press.

Rorty, R. (1986). Pragmatism, Davidson and truth. Reprinted in Rorty (1991b), pp. 126–150.

Rorty, R. (1991a). Introduction: antirepresentationalism, ethnocentrism and liberalism. In Rorty (1991b), pp. 1–17.

Rorty, R. (1991b). *Objectivity, Relativism, and Truth: Philosophical Papers Volume 1.* Cambridge: Cambridge University Press.

Rosenberg, A. (1985). *The Structure of Biological Science.* Cambridge: Cambridge University Press.

Roughgarden, J. (1979). *Theory of Population Genetics and Evolutionary Ecology: An Introduction.* New York: Macmillan.

Rumelhart, D. E., J. L. McClelland and the PDP Research Group (1986). *Parallel Distributed Processing: Explorations in the Microstructure of Cognition, Volume 1.* Cambridge MA: MIT Press.

Ruse, M. (1986). *Taking Darwin Seriously.* Oxford: Blackwell.

Russell, B. (1945). *A History of Western Philosophy.* New York: Simon and Schuster.

Salmon, W. (1989). *Four Decades of Scientific Explanation.* Minneapolis: University of Minnesota Press.

Saussure, F. de (1910). *Course in General Linguistics* (C. Bally and A. Sechehaye, ed., W. Baskin, trans.). New York: McGraw-Hill, 1966.

Schlichting, C. D. (1986). The evolution of phenotypic plasticity in plants. *Annual Review of Ecology and Systematics* 17: 667–93.

References

Schlick, M. (1932/33). Positivism and realism. Reprinted in Boyd, Gasper and Trout (1991), pp. 37–55.

Seger, J. and H. J Brockmann (1987). What is bet-hedging? *Oxford Surveys in Evolutionary Biology* 4: 181–211.

Selander, R. K. and D. W. Kaufman (1973). Genic variability and strategies of adaptation in animals. *Proceedings of the National Academy of Sciences, USA* 70: 1875–1877.

Shapin, S. (1982). History of science and its sociological reconstructions. *History of Science* 20: 157–211.

Shepard, R. (1987). Evolution of a mesh between principles of mind and regularities of the world. In Dupre (1987), pp. 251–275.

Simon, H. (1981). *The Sciences of the Artificial*. 2nd edition. Cambridge MA: MIT Press.

Skyrms, B. (1990). The value of knowledge. In C. W. Savage (ed.) *Minnesota Studies in the Philosophy of Science 14: Scientific Theories*. Minneapolis: University of Minnesota Press, pp. 245–266.

Smith, A. (1776). *An Inquiry into the Nature and Causes of the Wealth of Nations*. (E. Cannan, ed.) New York: Modern Library, 1937.

Sober, E. (1984). *The Nature of Selection*. Cambridge MA: MIT Press.

Sober, E. (1993). *The Philosophy of Biology*. Boulder: Westview.

Sober, E. (1994). The adaptive advantage of learning versus a priori prejudice. In *From a Biological Point of View*. Cambridge: Cambridge University Press.

Spencer, H. (1852a). The development hypothesis. Reprinted in *Essays Scientific, Political and Speculative*, Vol. 1. New York: Appleton, 1891, pp. 1–7.

Spencer, H. (1852b). A theory of population, deduced from the general law of animal fertility. *Westminster Review* 1: 468–501.

Spencer, H. (1855). *Principles of Psychology*. London: Longman, Brown and Green.

Spencer, H. (1857). Progress: its law and cause. Reprinted in *Essays Scientific, Political and Speculative*, Vol. 1. New York: Appleton, 1891, pp. 8–62.

Spencer, H. (1866). *Principles of Biology*. 2 volumes. New York: Appleton. (English edition 1864–66.) All references in the text are to volume 1.

Spencer, H. (1871). *Principles of Psychology*. 2nd edition. 2 volumes. New York: Appleton. (English edition 1870.) All references in the text are to volume 1.

Spencer, H. (1872). *First Principles of a New System of Philosophy*. 2nd edition. New York: Appleton.

Spencer, H. (1886). *Social Statics; or The Conditions Essential to Human Happiness Specified, and the First of Them Developed*. New York: Appelton (English edition 1850).

Spencer, H. (1887). *The Factors of Organic Evolution*. New York: Appleton.

Spencer, H. (1898). *Principles of Biology*. Volume 1. Revised and enlarged edition. London: Williams and Norgate.

Spencer, H. (1904). *An Autobiography*. 2 volumes. New York: Appleton.

306

References

Staddon, J. (1983). *Adaptive Behavior and Learning*. Cambridge: Cambridge University Press.

Stalnaker, R. (1984). *Inquiry*. Cambridge MA: MIT Press.

Stearns, S. C. (1989). The evolutionary significance of phenotypic plasticity. *BioScience* 39: 436–45.

Stephens, D. (1986). Variance and the value of information. *American Naturalist* 134: 128–140.

Stephens, D. (1991). Change, regularity and value in the evolution of animal learning. *Behavioral Ecology* 2: 77–89.

Sterelny, K. (1990). *The Representational Theory of Mind: An Introduction*. Oxford: Blackwell.

Sterelny, K. and P. S. Kitcher (1988). The return of the gene. *Journal of Philosophy* 85: 339–351.

Stich, S. P. (1990). *The Fragmentation of Reason*. Cambridge MA: MIT Press.

Stich, S. P. and T. A. Warfield (eds.) (1994). *Mental Representation*. Oxford: Blackwell.

Sultan, S. (1987). Evolutionary implications of phenotypic plasticity in plants, in *Evolutionary Biology*, Volume 21 (M. Hecht, B. Wallace and G. T. Prance, eds.) New York: Plenum, pp. 127–78.

Sultan, S. (1992). Phenotypic plasticity and the neo-Darwinian legacy. *Evolutionary Trends in Plants* 62: 61–71.

Sultan, S. E. and F. A. Bazzaz (1993). Phenotypic plasticity in *Polygonum persicaria*. III. The evolution of ecological breadth for nutrient environments. *Evolution* 47: 1050–1071.

Templeton, A. R. and E. D. Rothman (1974). Evolution in heterogeneous environments. *American Naturalist* 108: 409–428.

Templeton, A. R. and E. D. Rothman (1981). Evolution in fine-grained environments II: Habitat selection as a homeostatic mechanism. *Theoretical Population Biology* 19: 326–40.

Thayer, H. S. (1952). *The Logic of Pragmatism: An Examination of John Dewey's Logic*. New York: Humanities Press.

Thoday, J. M. (1953). Components of fitness. *Symposia of the Society of Experimental Biology* 7: 96–113.

Tierney, A. J. (1986). The evolution of learned and innate behavior: contributions from genetics and neurobiology to a theory of behavioral evolution. *Animal Learning and Behavior* 4: 339–348.

Tiles, J. E. (1988). *Dewey*. London: Routledge and Kegan Paul.

Tinbergen, N. (1963). On the aims and methods of ethology. *Zeitschrift für Tierpsychologie* 20: 410–33.

Todd, P. M. and G. F. Miller (1991). Exploring adaptive agency II: simulating the evolution of associative learning. In Meyer and Wilson (1991), pp. 306–15.

Van Fraassen, B. (1980). *The Scientific Image*. Oxford: Clarendon Press.

Via, S. (1987). Genetic constraints on the evolution of phenotypic plasticity. In V.

References

Loeschke (ed.) *Genetic Constraints on Adaptive Evolution*. Berlin: Springer-Verlag, pp. 47–71.

Waddington, C. H. (1957a). Discussion of Lewontin (1957). *Cold Spring Harbour Symposium on Quantitative Biology* 22: 408.

Waddington, C. H. (1957b). *The Strategy of the Genes*. New York: Macmillan.

Waddington, C. H. (1975). *The Evolution of an Evolutionist*. Ithaca: Cornell University Press.

Wade, M. (1978). A critical review of the models of group selection. *Quarterly Review of Biology* 53: 101–114.

Wallace, B. (1968). Polymorphism, population size, and genetic load. In R. C. Lewontin (ed.) *Population Biology and Evolution*. Syracuse: Syracuse University Press, pp. 87–108.

Waters, C. K. (1991). Tempered realism about the force of selection. *Philosophy of Science* 58: 553–573.

Webster, G. and B. C. Goodwin (1982). The origin of species: a structuralist approach. *Journal of Social Biological Structures* 5: 15–47.

Werren, J. H. (1980). Sex ratio adaptations to local mate competition in a parasitic wasp. *Science* 208: 1157–1159.

West-Eberhard, M. J. (1989). Phenotypic plasticity and the origins of diversity. *Annual Review of Ecology and Systematics* 20: 249–278.

West-Eberhard, M. J. (1992). Adaptation: current usages. In Keller and Lloyd (1992), pp. 13–18.

Whyte, J. (1990). Success semantics. *Analysis* 50: 149–157.

Wiener, J. (1949). *Evolution and the Founders of Pragmatism*. Cambridge MA: Harvard University Press.

Williams, G. C. (1966). *Adaptation and Natural Selection*. Princeton: Princeton University Press.

Wilson, D. S. (1983). The group selection controversy: History and current status. *Annual Review of Ecology and Systematics* 14: 159–87.

Wilson, D. S. and J. Yoshimura (1994). On the co-existence of specialists and generalists. *American Naturalist* 144: 692–707.

Wimsatt, W. (1972). Teleology and the logical structure of function statements. *Studies in the History and Philosophy of Science* 3: 1–80.

Winther, R. (in preparation). Darwin's views on environmental effects on organic variation.

Woolgar, S. (1988). *Science: The Very Idea*. London: Tavistock.

Wright, L. (1973). Functions. *Philosophical Review* 82: 139–168.

Wright, L. (1976). *Teleological Explanations*. Berkeley: University of California Press.

Young, R. M. (1970). *Mind, Brain and Adaptation in the Nineteenth Century*. Oxford: Clarendon.

Index

Index

Papineau, D., 6, 167, 168, 176, 180, 182–3, 201n17, 201n18
Peel, J. D. Y., 69, 94n1, 96n10, 98n23
Peirce, C. S., 6, 14, 128n2, 138, 193–4
phenotypic plasticity, 208–9, 214–15, 229n2
Piatelli-Palmerini, M., 87
Popper, K. R., 25, 33, 98n25, 99n28, 115, 126, 189
Piaget, J., 52, 60n1, 64n20, 65n30, 72, 80
polymorphism, genetic, 255–7
  as consequence of heterosis, 257–8
  as consequence of variation in selection, 264–87
  related to homeostasis, 258–63
pragmatism, 4, 6–7, 14, 92, 107, 110, 117–20, 138, 167–71, 192–5, 196
Putnam, H., 6, 62n6, 64n24, 193

quasi-homeostasis, 269–70, 284–5
Quine, W. V., 167, 196nl

Ramsey, F. P., 14, 168, 179, 200n13
Ramsey, W., 201n20
reductionism, 31, 42–3, 44
Reeve, E. C. R., 260–1
reliabilist epistemology, 3, 36, 116–20
reliability, 232–38, 247–53
Richardson, R., 61n1
Robertson, F. W., 260–1
Rorty, R., 100, 127n1, 167–8, 194
Rosenberg, A., 189
Rothman, E. D., 267–8
Roughgarden, J., 189, 272, 282, 288n13, 290n25
Rumelhart, D., 28, 61n4
Ruse, M., 94n4, 98n22, 99n28

Salmon, W., 17, 97n18
Schlick, M., 34
Seger, J., 213, 279, 290n27
selective mechanisms, 16, 86–90, 92–4, 98n27, 113–16, 126–7, 197n3
Shapin, S., 63n14, 162
Shepard, R., 35
signal detection theory, 209, 212, 232–8, 286
Simon, H., 29n10, 36, 48, 64n27

Skyrms, 217, 230n9
sociology of science, 41, 63n14, 161–2
Sober, E., 54, 65n32, 190, 191, 203n25, 209, 229n5, 288n8
Spencer, H., 5–7, 9, 35, 58, 66–99 passim, 138–9
  on correspondence, 166–9, 197–8
  on Darwin, 93–4
  on evolution, 79–84
  on life, 69–72
  on mind, 72–6
  on progress, 139–40
  on psychology, 84–90
Staddon, J., 65n33, 213
Stalnaker, R., 129n7, 177
Stearns, S., 209
Stephens, D., 220–1, 229n3, 230n10
Sterelny, K., 62n6, 175, 176, 264
Stich, S., 129n7, 168, 171, 173, 175, 194, 199n6, 199n10
Sultan, S., 229n2, 239, 244, 253n4, 289n20

teleonomic
  sense of function, 16–20, 191
  theories of meaning, 180–3
version of environmental complexity thesis, 21–4, 59–60, 90, 109–13, 124–5, 195
Templeton, A. R., 267–8
Thayer, H. S., 140, 165n6
todd, P. M., 121–4, 216
trial and error, see selective mechanisms

Van Fraassen, B., 34
Varela, F., 49, 50, 96n13
Via, S., 209, 229n2

Waddington, C. H., 52, 65n30, 78, 288n7
Warfield, T., 129n7, 175
Whyte. J., 179, 200n14
Williams, G. C., 16, 253n7, 263
Wilson, D. S., 288n8, 289n18
Wimsatt, W., 17
Winther, R., 164n2, 98n24
Woolgar, S., 41, 161–3, 165n11
Wright, L., 15–6, 28n1, 28n2, 191

311

Printed in the United States
By Bookmasters